DESIRE NZOYISABA
BENOIT NZIGIDAHERA
BERNADETTE HABONIMANA

Vegetation survey of Kibira National Park: Teza sector

DESIRE NZOYISABA
BENOIT NZIGIDAHERA
BERNADETTE HABONIMANA

Vegetation survey of Kibira National Park: Teza sector

ScienciaScripts

Imprint

Any brand names and product names mentioned in this book are subject to trademark, brand or patent protection and are trademarks or registered trademarks of their respective holders. The use of brand names, product names, common names, trade names, product descriptions etc. even without a particular marking in this work is in no way to be construed to mean that such names may be regarded as unrestricted in respect of trademark and brand protection legislation and could thus be used by anyone.

Cover image: www.ingimage.com

This book is a translation from the original published under ISBN 978-620-3-44961-7.

Publisher:
Sciencia Scripts
is a trademark of
Dodo Books Indian Ocean Ltd. and OmniScriptum S.R.L publishing group

120 High Road, East Finchley, London, N2 9ED, United Kingdom
Str. Armeneasca 28/1, office 1, Chisinau MD-2012, Republic of Moldova, Europe
Printed at: see last page
ISBN: 978-620-8-06278-1

DEDICATION

To my parents and relatives;
To my brothers and sisters;
To my beloved Maniragumije Aline; To my cousins;
To my friends and friends of nature; I dedicate this memoir.

ACKNOWLEDGEMENTS

As this work draws to a close, we would like to take this opportunity to express our sincere thanks to all those who have contributed to its completion.

Our first sincere thanks go to Mr. Benoît NZIGIDAHERA, in charge of research at OBPE (formerly INECN) and Professor Bernadette HABONIMANA, lecturer at the Faculty of Agronomic Sciences of the University of Burundi, respectively director and co-director of this thesis, who, despite their multiple occupations, agreed to supervise this work. Their scientific rigour, their advice and their frank collaboration have been of great use to us.

We would also like to thank the managers of OBPE (formerly INECN) and in particular Mr. Jean Claude HAKIZIMANA, head of the Kibira National Park, as well as the forest rangers of the Teza sector for their open collaboration during fieldwork. We would also like to take this opportunity to express our gratitude to Mr. Blaise NDIHOKUBWAYO and Barnabé NIZIGIYIMANA for looking after us while we were in the field.

We'd also like to thank all our family members, who put up with the cost of our time spent on the school bench. In particular, we would like to thank the NTIBAGIRIRWA Gaspard family for taking care of our stay in Bujumbura before we entered and after we left the university residences.

I would like to express my satisfaction to all the teachers who took part in our training program from elementary school to the University of Burundi.

Last but not least, we would like to thank the Royal Belgian Institute of Natural Sciences (IRScNB) for funding the fieldwork for this project.

Nzoyisaba Désiré

ACRONYMS AND ABBREVIATIONS

A-AM   Arborescent with small to medium-sized trees

aB Shrubbery

A-GA   Arborescent with large trees

A-TGA   Arborescent with very large trees

FACAGRO   Faculty of Agronomic Sciences

GPS   Global Positioning System

Ht Height

IGEBU Institut Géographique du Burundi

INCN   National Institute for Nature Conservation

Indét   Indeterminate

INECN Institut National pour l'Environnement et la Conservation de la
Nature

IRScNB Royal Belgian Institute of Natural Sciences

Lat Latitude

LEM   Law Enforcement Monitoring

Lon Longitude

Nbr Number

NNW   North North West

N-S North-South

OBPE   Office Burundais pour la Protection de l'Environnement (Burundian Office for
Environmental Protection)

O.R.U   Rwanda-Urundi Ordinance

KNP   Kibira National Park

Rc Recovery

SsAH   Sub-shrub and/or herbaceous

SSE South South -East

Topo   Topography

IUCN   International Union for Conservation of Nature

TABLE OF CONTENTS

INTRODUCTION

Burundi's position in the center of Africa, its topography, its territory combining both farmland and aquatic land, and its diversity of eco-climatic conditions give it an incredible wealth of plant and animal species and diversified forest ecosystems (Habonimana *et al.*, 2007). Currently, Burundi has 17 protected areas in 4 categories, including 3 national parks, 8 nature reserves, 1 natural monument and 5 protected landscapes (Masharabu, 2012). The same author points out that all Burundi's protected areas have a surface area of around 13,6706.48 ha, or 5% of the territory. Among the 3 national parks is the Kibira National Park (PNK), which has had the privilege of being selected as one of the model forests that could serve as an example for other forests in the sub-region as part of the conservation and sustainable management of Central African tropical rainforest ecosystems (FAO, 2002). It is, in fact, one of the most appropriate forest reserves for mountain forest conservation.

The Kibira National Park, which covers 40,000 ha, offers undeniable advantages, not only in terms of cleaning up the atmosphere like any other forest ecosystem, but also as a major water tower for a large part of Burundi. Indeed, major rivers such as Kaburantwa, Kagunuzi, Mpanda, Ruvubu and Gitenge originate here (IUCN, 2011). It contains over 644 plant species, including those used by man to satisfy his many needs (timber and service wood, medicinal plants and edible and non-edible non-timber forest products) (Nzigidahera, 2000).

Despite its vital importance for Burundi and other countries in the region, reflected in the multiplicity of its goods and services, Kibira National Park is one of the most threatened forest environments (Habonayo and Ndihokubwayo, 2011). Current mountain forest vegetation is strongly influenced by anthropization (Nzigidahera, 2012). High densities in areas bordering Kibira, poverty and underdevelopment of the surrounding population, contribute to a progressive degradation of the Park's resources.

Traditional uses in the Kibira include deadwood harvesting, livewood harvesting, litter harvesting, bamboo cutting, traditional pharmacopoeia, marsh grass harvesting, honey harvesting and hunting (FAO, 2002). Species already reported as endangered include *Entandrophragma excelsum, Prunus africana,*

Symphonia globulifera and *Hagenia abyssinica* (Nzigidahera, 2000; Habonimana *et al.*, 2007).

To mitigate these threats, Kibira National Park is guarded by a corps of forest rangers whose main role is to track down criminals. This system provides no information on the evolution of the protected ecosystems. The reports produced provide no information on the dynamics of the Park's biodiversity. The absence of continuous monitoring of biodiversity means that it is impossible to say exactly how fast animal and plant species are disappearing in Burundi. OBPE (formerly INECN), in collaboration with Belgium's Institut Royale des Sciences Naturelles, would like to set up a solid system for monitoring populations and species in Burundi's national parks. To help set up this monitoring system, a study entitled **"Etude de la végétation du Parc National de la Kibira: *Secteur Teza*"** has been carried out.

The overall aim of this study is to build up a database needed to monitor the dynamics of habitats, populations and species in the management of Kibira National Park.

The specific objectives are:

- Carry out a floristic inventory of the various work sites;
- Analyze vegetation physiognomy at different sites;
- Characterize the phytogeographical features and biological forms of the species collected;
- Analyze stand density at the various sites surveyed;
- Quantify litter and calculate basal area;
- Establish the relationship between litter, basal area and wood density;
- Highlight the evolutionary state of habitats and guide the monitoring of their dynamics.

In this work, we are guided by the following hypotheses:

- The habitats of the Kibira National Park (Teza sector) are located at different levels of evolution;
- Floristic and structural characteristics vary from one habitat to another;
- The habitats of Kibira National Park are undergoing a regressive evolution.

Our work is divided into four chapters. The first chapter deals with the description of Kibira National Park, the second with general information on habitat dynamics, the third with the methodology used and the fourth with the results and their discussion.

I. DESCRIPTION OF KIBIRA NATIONAL PARK

I.1. LOCATION

Kibira National Park covers 40,000 ha, is 80 km long and never more than 8 km wide. Located in north-western Burundi, it is the country's main forest massif. It occupies the summit of the Congo-Nile ridge, which extends into Rwanda as far as Lake Kivu, taking the name of Nyungwe (Nderagakura, 1999). The KNP straddles four provinces of Burundi: Bubanza and Cibitoke to the west, Muramvya and Kayanza to the east. The KNP lies between 2^0 26'52" and 3^0 17'o8" south latitude and between 29^0 13'31" and 29^0 39'09" east longitude. Its altitude varies between 1,600 and 2,666 m (highest point of the Congo-Nile ridge).

According to Gourlet (1986), KNP is divided into four blocks or sectors:

- Secteur Teza: 5,794 ha (Muramvya)
- Musigati sector: 15,424 ha (Bubanza)
- Rwegura sector: 12,423 ha (Kayanza)
- Mabayi sector: 6359 (Cibitoke)

Our study area is the Teza sector (Fig.1).

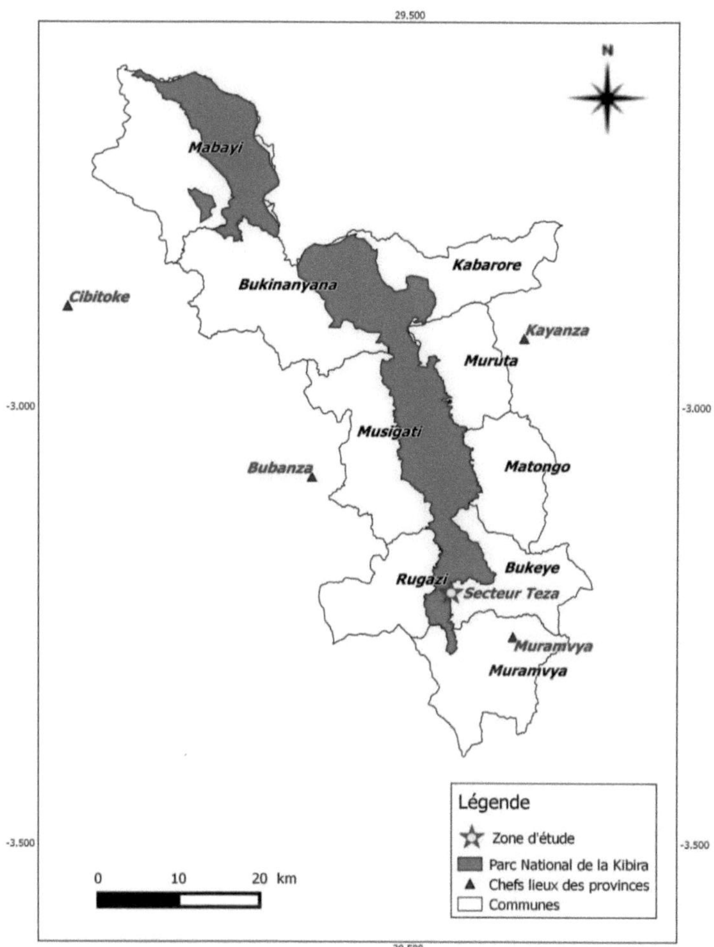

Fig. 1: Location of Kibira National Park (Author)

I.2. PHYSICAL FRAME

I.2.1. Climate

KNP has a temperate, high-altitude tropical climate, marked by its mountainous character. Average temperatures vary between 11.7 and 21° depending on altitude.

Over the last 30 years (1984-2014), the average monthly temperature recorded at the Rwegura weather station is 16.4°c. June and July record the lowest temperatures (15.85°), while September is the hottest month of the year, with an average of 17.05°c.

The average annual rainfall recorded at the Bugarama and Rwegura weather stations is 1,627 mm. April has the highest rainfall (222.8 mm), while July has the lowest (6.7 mm) (IGEBU, 2015). Figures 2 and 3 respectively show the umbrothermal diagram and the water balance for KNP constructed from monthly averages for the 1984-2014 period from the Bugarama and Rwegura weather stations.

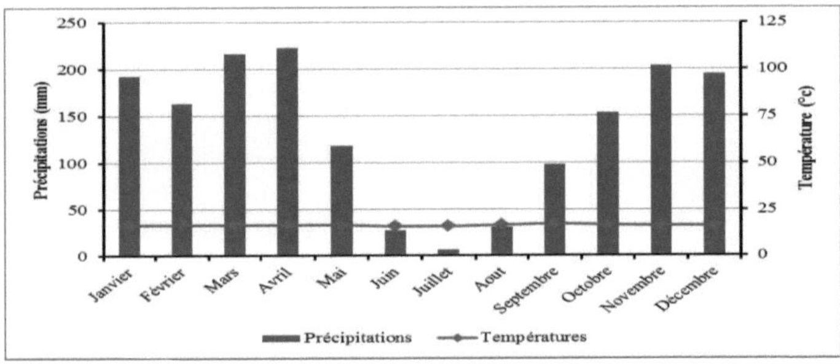

Fig. 2: Umbrothermal diagram for KNP constructed by the author using the Gaussen model and averages for the period 1984-2014 from the Bugarama and Rwegura weather stations.

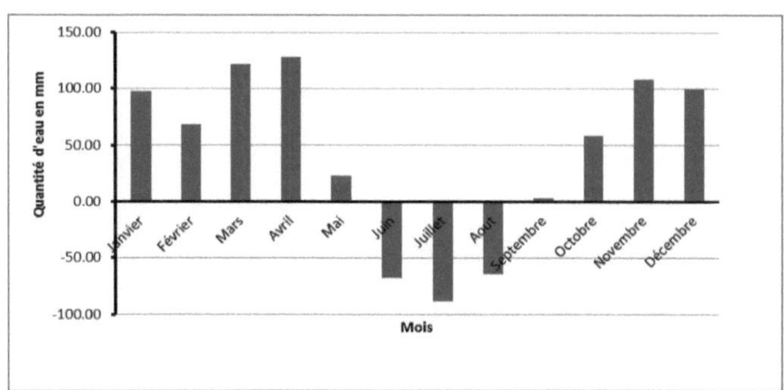

Fig.3: Water balance for KNP constructed by the author using the Gaussen model and averages for the period 1984-2014 from the Bugarama and Rwegura weather stations.

I.2.2.   Hydrography

The KNP is considered Burundi's water tower. Many rivers have their source in this forest massif. The ridge line delimits two watersheds commonly referred to as the "Nile Basin" to the east and the "Congo Basin" to the west. The streams and rivers to the west of KNP flow into the Rusizi (Habonayo and Ndihokubwayo, 2011). From south to north, the main ones are the Ruhora, Mpanda, Gitenge and Kaburantwa rivers.

The streams and rivers to the east of KNP flow towards the central plateau (Ruvubu) and Bugesera (Kanyaru). The main ones are: Nyabibondo, Nkoma, Nyakabimbi, Kayave and Buyumpu (Habonayo and Ndihokubwayo, 2011).

I.2.3.   Topography and geomorphology

The whole of this mountain ridge forms a very open arc stretching in a SSE/NNW direction between Bugarama in Burundi and the Rwandan border. The massif continues further south, in a much more subdued form and parallel to the Lake Tanganyika Graben.

The KNP is relatively narrow overall, but widens between the Rwandan border and Rwegura. Its relief is marked by steep slopes on both sides of the Congo-Nile ridge, more pronounced on the western side (Nderagakura, 1999).

The highest points from south to north are Teza (2666 m), Musumba (2661 m), Dusasa (2621 m) and Twinyoni (2559 m) (Lewalle, 1972).

I.2.4. Pedology

The pedological development of Kibira soils is characterized by the presence of a humus-bearing horizon linked to the presence of the rainforest. This horizon, which is sometimes very deep, has characteristics closely linked to the mountain climate. It consists of a litter layer (2 to 4 cm thick) overlying a black humus horizon highly saturated with cations, and is finely lumpy, loose to crumbly in structure (Lewalle, 1972).

The upper horizon is humus-bearing, but the absorbing complex is highly saturated and its structure is more massive. Everywhere and at varying depths, the humus-bearing horizon presses seamlessly into a massive horizon of very low bulk density. Mountain forest soils generally have very good fertility potential. They are heavy clay Ferrisols with local intrusion of schists or basic rocks, and clayey Humiferous Ferralsols (Lewalle, 1972).

Two soil zones have been identified in Kibira: north of Teza consists of poor-quality, undifferentiated lithosols; in the Teza region, rich andosols are found (Cazenave-Piarrot *et al.*,1979 cited by Habonayo and Ndihokubwayo, 2011).

I.2.5. Geology

The multiple lithological facies can be grouped into four major groups (INECN, 1998 cited by Wibereho, 2010):

- **Quartzite ridges**

These ridges are composed of a variety of hard rocks, dominated by quartzites. The slow rate of weathering of these rocks, combined with erosion due to the steep slopes (30-60%), means that lithosols and regosols are predominant. The most that can be observed is a slight deepening of the soils through colluvium in the small talwegs or any flat areas.

- **Granite and granito-gneiss**

Granites and granito-gneisses are acidic rocks that weather to produce a smaller quantity of clay than basic rocks. The quartz fraction, which is practically unalterable, releases inert sands that remain in place. The feldspars, on the other hand, will alter, all the faster for being low in silica, and transform into clay minerals that can migrate vertically and laterally. Soils on granites and granito-gneiss are coarse-textured and chemically poor.

- Schisto-metamorphic complex

Backed by quartzite or granite-gneissic ridges, it is formed by structural advances with shallower slopes, locally evolving into rounded-top hills. Although irregular, weathering is generally significant on the micaschistose formations and soils are deep. Clay content, at around 50%, is fairly regular across the profile, with Kaolinite clearly dominant. The silty fractions are composed of phyllosilicates (kaolinite, muscovite) and quartz. Quartz and muscovite are the only elements in the sandy fraction.

- Alluvium

They are only found in the Kibira where a rocky sill blocks a valley with a gentle longitudinal gradient. They are mainly found in the upper Gitenge and Mpanda valleys (Ntibarirarana, 2002).

I.3. BIODIVERSITY OF KIBIRA NATIONAL PARK

I.3.1. Vegetation

This rugged environment is mainly dominated by mountain rainforest, at altitudes ranging from 1,600 to 2,500 m (Niyukuri, 2012). Overall, over 644 plant species are already known to exist in KNP (Nzigidahera, 2007). According to Lebrun (1935), quoted by Lewalle (1972), three horizons can be distinguished in the mountain rainforest based on altitudinal limits and physiognomic and floristic criteria:

- Lower horizon (1600-1900 m)

A characteristic physiognomic feature of lower-horizon forests is the high density and intertwining of tree strata. In the upper tree stratum, powerful trees such as *Anthonotha pynaertii, Albizia gummifera, Parinari excelsa, Prunus africana and Syzygium staudtii* can reach 25 m in height, with dense, broad crowns.

The lower arborescent stratum is very varied; locally, *Carapa grandiflora* is abundant. Large lianas, *Securida cawelwitschii, Schefflera barteri*, encumber both tree strata. The undergrowth of shrubs and suffrutex is almost impenetrable in places.

In all strata, epiphytes are abundant and varied, especially ferns and orchids, in addition to numerous mosses and liverworts.

- Middle horizon (1900-2250 m)

In the montane rainforest of the middle horizon, the strata are quite distinct. The upper tree stratum, reaching 30 m and sometimes 40 m, is made up of giants such as *Entandrophragma excelsum, Prunus africana, Parinari excelsa* and a few secondary species, notably *Polyscias fulva*. The dominant tree stratum is rich in species: *Tabernaemontana johnstonii, Symphonia globulifera, Strombosia scheffleri* and numerous secondary species, *Xymalos monospora, Bersama ugandensis, Macaranga neomildbraediana, Neoboutonia macrocalyx,* etc. In the shrub stratum, *Dracaena afromontana* is particularly frequent, along with *Galiniera coffeoides, Allophylus oreophilus, Rauvolfia obscura, Chasalia subochreata*, and others. Creepers climb to the top: *Coccinia mildbraedii, Jasminum pauciflorum, Culcasia scandens*. Typical forest grasses include *Oplismenus hirtellus, pseudechinolaena polystachya*, Ferns, Balsamines, etc. Epiphytes are very abundant, covering trunks and stumps as well as high branches. These are mainly ferns and Lycopodes, with orchids less numerous than in the lower horizon.

- Upper horizon (2250-2450 m)

The appearance of the forest on this horizon is quite distinct from that on the middle horizon. Apart from a few exceptional *Podocarpus milanjianus* trees reaching 20 m, the crown generally stops at 15 m. The lower tree stratum is sparsely populated.

The shrub layer is made up of special species: *Monanthotaxis orophila, Maytenus acuminatus, Rapanea pulchra*. Most branches are covered with epiphytes, but these are rarely ferns or orchids, but rather mosses and lichens. Usnea appear in the brightest parts. The herbaceous layer is discontinuous and species-poor.

In addition to the natural indigenous vegetation, various exotic species were planted in buffer plantation blocks covering some 2,000 ha, and 49 ha were planted to enrich the forest with *Entandrophragma excelsum*.

I.3.2. Fauna

According to IUCN (2011), the main mammals found in KNP are: Harnessed guib (*Tragelaphus scriptus,* Bovidae), bushpig (*Potamochoerus larvatus,* Suidae), yellow-backed duiker (*Cephalophus silvicultor*, Bovidae), black-fronted duiker (*Cephalophus nigrifrons*, Bovidae), serval (*Leptailurus serval,* Felidae), the striped jackal *(Canis adustus,* Canidae), the civet cat *(Civettictis civetta,* Viverridae) and a wide variety of primates such as the diademed cercopithecus *(Cercopithecus mitis dogetti,* Cercopithecidae) and the chimpanzee (*Pan troglodytes*, Homnidae). There are also 20 species of insectivores, some of which are endemic. The avifauna is highly diverse, with over 200 species.

I.4. PROTECTING THE KIBIRA NATIONAL PARK

I.4.1. History of its creation

The history of KNP can be divided into 6 periods
(Nzigidahera, 2000):

-   Before 1933: Kibira was the forest used as a hunting reserve by the kings of Burundi. Local populations respect the forest, which they consider to have a magical function. Local chiefdoms may allocate land for farming. A right of use is recognized for grazing and gathering forest products (timber and firewood, pharmacopoeia, beekeeping, bamboo and marsh grasses, etc.). The Batwa know the forest and live off its products;

-   Period from 1933 to 1962: under Belgian trusteeship, the KNP was classified as a forest reserve by the Belgian authorities under O.R.U n° 33/agri. of 24/05/1934. A double line of cypress trees was planted to demarcate the park. Around 1960, pressure on the Kibira increased due to the high density of the surrounding population (around 300 inhabitants/km^2). Access to new land in the Kibira was progressively forbidden in the face of the threat of its total clearing for agriculture;

- 1962 to 1980: from Burundi's independence in 1962, the reserve was administered by the Burundian authorities (Water and Forestry Department). At that time, only the exploitation of valuable woods (mainly *Entandrophragma excelsum and Prunus africana)* was regulated and controlled. Within the demarcated perimeter, the right to allocate new land for cultivation was abolished. The right to use grazing land was retained; those concerning other products were tolerated. Many farmers were able to cross the boundaries, clear land and cultivate without being bothered, because the personnel responsible for enforcing regulations and monitoring the boundaries were few and it was difficult to gain access to certain parts of the massif;

- From 1980 to 1993: the Institut National pour la Conservation de la Nature (INCN) was created to administer the protected areas, under the direct supervision of the President of the Republic. From this point onwards, Kibira was declared a National Park. Use rights were no longer tolerated within the perimeter of the Park. Grazing, grazing fires and the collection of products other than dead wood were prohibited. A double line of pine trees was planted, a perimeter surveillance trail was opened and plantations were planted to protect and restore the soil;

- 1993 to 2000: since 1993, laws and regulations have been disregarded as a result of the country's insecurity. Local authorities were no longer listened to or respected. Some local residents take advantage of the situation to cross boundaries, clear land and plant crops, cut down large, valuable trees, burn steppes and forests, and destroy plantations, tracks and shelters;

- 2000 to 2008: in 2000, a legal status for KNP was established through decree no.° 100/007 of January 25, 2000, delimiting a national park and four nature reserves. However, protection of the Park has not improved. The war situation has led to massive destruction at Bugarama, and a number of private individuals have been given land in the Park for farming and livestock rearing.

In 2004, these private individuals abandoned part of the Kibira land in Bugarama.

A system of co-management of protected areas by the State and communities was introduced by INECN in 2008.

INECN became the Office Burundais pour la Protection de l'Environnement (OBPE) in 2014 by decree-law no.° 100/240 of October 29, 2014.

I.4.2. KNP attributes

I.4.2.1. Ecological attributes

The Kibira mountain rainforest plays a fundamental role in regulating the water regime and protecting watersheds on steep slopes from erosion.

Indeed, it is in the Kibira region that Burundi records its highest rainfall. Many rivers have their source in this forest (Nzigidahera 2000). The forest provides many services and influences that are highly beneficial to environmental protection. It plays an essential role in regulating climate and atmospheric composition. It lowers temperatures, increases rainfall and attcnuates mctcorological fluctuations (Nderagakura, 1999).

Mountain forests also provide the essential conditions for the perpetuation of a wide variety of biological species, many of them endemic. They are home to plants that serve as food for wild animals and riverrains. The rainforest is also the habitat of choice for endangered animal species such as *Pantroglodytes schweinfurthii* (Nzigidahera, 2000).

I.4.2.2. Socio-economic attributes

The KNP is a supply point for the local and remote population. It offers food of high nutritional value, as well as preventive and curative medicines appreciated by the population. Boards and planks from the Kibira provide considerable cash income for the farmers (Ndayikeza and Niyimpaye, 2007).

The Kibira forest provides essential hydrological and climatic conditions for the country's agriculture. In fact, it helps to prevent floods, landslides and landslides in the interests of hill and marsh farming. The Kibira is a source of water for farmland in large parts of the country. Much of northern Burundi is irrigated by water from the Kibira. The KNP offers a favorable climate for tea-growing. There are 3 tea factories close to the Kibira: Teza, Rwegura and Buhoro. Power generation and associated economic development are strongly linked to the Kibira mountain forest, which feeds and maintains the flow of the dams. The country's major hydroelectric dam at Rwegura on the Gitenge currently supplies 50% of the country's electricity needs (Nzigidahera, 2000).

I.4.3. Factors hindering protection of KFN

The KNP is protected for the role it plays in soil stabilization, water and climate regulation and the preservation of its flora and fauna (Nshimirimana, 1994). Although Kibira has been protected since 1933, various forms of exploitation of the forest's natural resources have long contributed to its degradation. The KNP's surrounding population is both large and unevenly distributed, accentuating its impact on the park's resources. The highest densities mean insufficient land for cultivation, leading to increased recourse to forest clearing. The high number of households increases the need for firewood, for which the Kibira is often described as an inexhaustible source (Nzigidahera, 2000).

Illegal cutting of trees and other products has been observed, which the surrounding population uses for pharmacopoeia, basketry and the manufacture of all kinds of tools such as doors, pirogues, baskets, etc. (Ndayikeza and Niyimpaye, 2007).

Nzigidahera (2007) reports that *Arundinaria alpina* bamboo is one of the most sought-after plant resources, particularly in Kayanza province. It is one of the most important non-timber forest products used for a variety of purposes (house building, basketry, furniture, cultural instruments, etc.). Hunting also harms the park's wildlife. Cattle breeders seek pasture in the Kibira, and beekeepers set up hives there, often causing fires during the honey harvest.

II. GENERAL INFORMATION ON HABITAT DYNAMICS

II.1.   DEFINITION AND OBJECTIVES

According to Salvaudon (2006), vegetation dynamics is the study of changes in vegetation over time. It ranges from very short periods (seasonal changes) to much longer periods (vegetation history).

The same author states that the objectives of monitoring vegetation dynamics are as follows:

- Orient   the   management (whether whether conservation,   economic, landscape, etc.);
- Improve knowledge of vegetation dynamics in a given area site;
- Acquire direct heritage information (monitoring particular plant species, natural habitats);
- Aquire   from   information   heritage   indirect   (on the
animals, etc.);
- Follow an experimental set-up.

II.2.   HABITAT SUCCESSION

A fundamental characteristic of ecological systems is their dynamism. Even superficial observation shows that bare soil gradually becomes covered with vegetation, and that an abandoned field is gradually invaded by grasses, perennials, then shrubs and finally trees (Guinochet, 1973 cited by Sedjar, 2012). So the natural dynamics of plant groupings generally move from simple to complex structures. This phenomenon of colonization of an environment by living beings and change of flora over time is referred to as "succession" (Sedjar, 2012). This ecological succession will continue for decades or even centuries until it reaches its ultimate stage of evolution, known as climax (Ramade, 2009). According to the same author, climax designates a stable association of species that characterizes qualitatively and quantitatively the ultimate development phase of a biocoenosis in a succession.

Based on photo-interpretation and observations made in the Kibira forest, a degradation process has been established (Table 1). By eliminating large trees from the upper strata, the mountain rainforest of the lower and middle horizons (*Entandrophragma excelsum* and *Parinari excelsa* ssp.*holstii* rainforest) becomes a secondary forest with *Polyscias fulva* and *Macaranga kilimandscharica*. Considerable impoverishment of the lower stratum of the upper-horizon montane secondary forest (secondary forest with *Syzygium parvifolium* and *Macaranga kilimandscharica*) gives rise to a secondary forest with *Faurea saligna*. Removal of the shrubs will result in a herbaceous plant formation. Degradation of the Ericaceae fruticée is evidenced by the disappearance of the shrub layer, giving way to altimontane grassland. With further degradation, and due to erosion, the substrate appears and the rock outcrops. Once the vegetation has degraded, when environmental conditions return to normal, the recolonization process is triggered. In the lower and middle montane horizons, recolonization begins with a herbaceous stratum of *Pteridium aquilinum*, which evolves into secondary forest. In the upper montane horizon, on the most exposed ridges where erosion has taken place and the soil has become superficial, the degraded low meadow formation may evolve into a recolonizing Ericaceae fruticée. The process may continue until secondary *Faurea saligna* forests are established. In a rich, uneroded environment, the recolonization formation evolves into a secondary forest with *Syzygium parvifolium* and *Faurea saligna*. In the afro-subalpine zone, the recolonization of the altimontane grassland takes place with species such as *Protea madiensis, Agauria salicifolia, Erica benguellensis* in areas where the rock outcrops, *Hypericum revolutum, Faurea saligna* in places where the soil is still present. Habiyaremye (1995) describes the same evolutionary process in mountain forests (Fig.4). According to this author, the process from post-cultivation to pre-forest recrûs is quite long. The phase from pre-forest recrûs to secondary forest is very rapid, as the elements of these stages are simultaneously present on the ground. Under the mantle of secondary forest, plants from the specific core of dense forest find ideal conditions for their growth (shade, humidity, etc.).

Overall, a dense climax forest matures in 40 years.
(Habiyaremye, 1995).

Table 1: Regressive evolution in montane rainforest (established by Nzigidahera, 2012 based on data from Gourlet,1986)

Degradation process: Afro-montane stage	
Removal of large trees from upper strata	• Mountain rainforest of the lower and middle horizons (*Entandrophragma excelsum* and *Parinari excelsa* rainforest). ↓ • Secondary forest with *Polyscias fulva* and *Macaranga kilimandscharica.*
Considerable impoverishment of the lower stratum	• Upper-horizon montane secondary forest (secondary forest with *Syzygium parvifolium* and *Macaranga kilimandscharica*) ↓ • *Faurea saligna* secondary forest ↓ • Herbaceous plant formation
Degradation process: Afro-subalpine stage	
The disappearance of the shrub layer	• Ericaceae Fruticaceae ↓ • Altimontane meadow ↓ • Substrate and rock outcrop

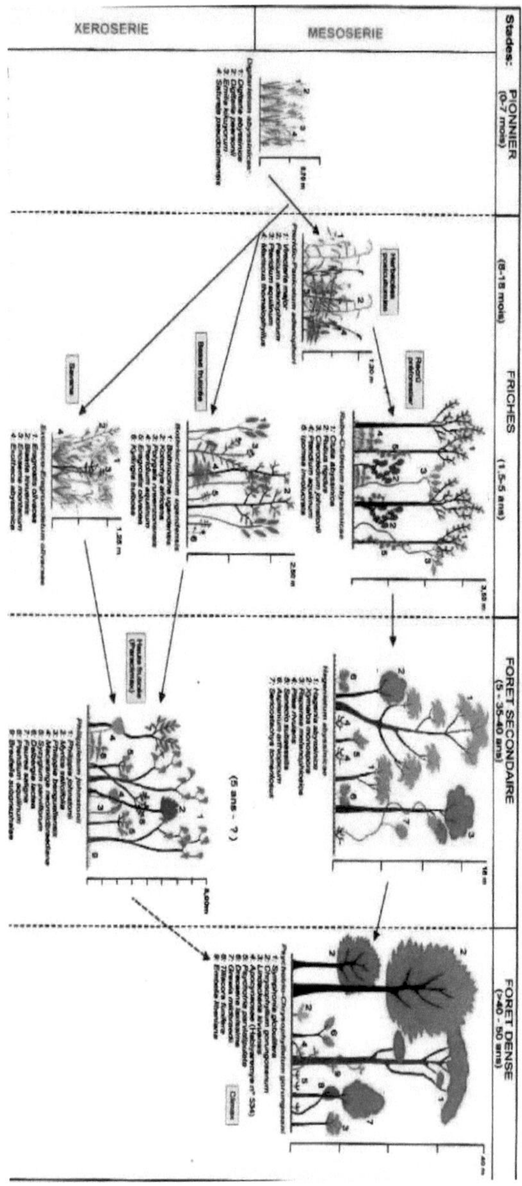

Fig. 4: Coexisting phytodynamic stages on the Congo-Nile ridge (Habiyaremye, 1995)

II.3.   MAIN FACTORS IN THE EVOLUTION OF HABITATS

The main factors behind evolution, especially regressive evolution, are of two kinds: anthropogenic factors and natural factors. Anthropogenic factors represent the major disturbance and regression factors in dynamic series (Sedjar, 2012). Indeed, man acts through the removal of firewood, felling of timber, clearing, cutting, fires, etc.. Natural factors include climate change, intensification of erosive processes and the impact of certain animal populations (such as phytophagous insects). For progressive evolution, it is only the maintenance of ideal conditions that ensures this evolution (Manirakiza, 2013).

II.4.   MONITORING HABITAT DYNAMICS

The aim of all methods for monitoring vegetation dynamics is to save time in understanding the environment. They aim to establish standard data that can be compared and processed. The best methods are those that deliver tangible results as quickly as possible, while optimizing the time spent collecting and processing information (Salvaudon, 2006). These methods can be divided into two categories: indirect and direct.

- **Indirect methods**

They are used to determine past vegetation trends,
or for monitoring on a very large scale. Examples:

- Monitoring the evolution of plant formations by interpreting and comparing old and current aerial photos;
- Comparison of successive forest management documents
for monitoring forest stand types;
- Comparison of old and recent vegetation surveys at the scale of a given region, to assess global changes in flora.

- **Direct methods**

These are methods for directly monitoring changes in vegetation over a fixed, precise study area, with a study period adapted to the problem.

The essential points are the same in all cases:

- Good location of survey plots;
- A simple, clear description of the protocol to be implemented in order to
to avoid operator change bias;
- The best thing to do is to set up a statement form to be filled in with a
leaflet;
- To obtain results that can be used quantitatively, the number of readings
taken, the sampling plan and the method used to record the results must enable
statistical processing,...

In Burundi, a LEM (Law Enforcement Monitoring) sheet was drawn up by
Habiyaremye in 2012 to monitor habitat dynamics in the country's protected
areas (Appendix 4). The LEM sheet briefly describes the physiognomic aspect
of a habitat, identifying the dominant species in each stratum and their cover.
So, by analyzing a LEM sheet for a habitat, we can get an idea of its likely
evolution. In addition, by comparing LEM sheets drawn up at different times
for the same habitat, we can assess the changes observed.

III METHODOLOGY
III.1. MATERIALS USED

The equipment used is as follows:

-    Knives for cutting specimens for harvesting,
-    Serpettes for clearing the path and cutting the stakes used to demarcate the sites,
-    Newsprint and presses for transporting harvested specimens,
-    Nylon ropes for making quadrants,
-    Braisier for drying harvested specimens,
-    Plastic bags to protect specimens from rain,
-    Folders, cardboard, scissors for making a herbarium,
-    Digital camera for taking photos,
-    Decameter for site demarcation,
-    Tape measure for measuring circumferences,
-    GPS for geographic coordinates,
-    Suunto dendrometer for measuring the height of standing trees,
-    U-shaped metal rod, 4 mm in diameter (see fig. 6),
-    Graduated plate for measuring the size of litter elements.

III.2. SAMPLING METHODS

III.2.1. Tracing the ecological trail

The starting point for this study of habitat dynamics is the establishment of a permanent transect, considered as an ecological pathway along which the Park's biological diversity is observed. At the start of the work, a 1.11 km east-west transect was traced. This transect, which begins near the OTB tea plantations, corresponds to the transect previously established for chimpanzee monitoring.

It crosses several habitat types, which is why it was deemed interesting for our study. It is a permanent transect, as it will be used for several years to monitor the evolution of the forest. In addition, other scientific research can be carried out along the same transect.

Six sites described in Table 2 are defined along the transect on the basis of physiognomy and floristic homogeneity. The geographic coordinates taken at these sites were used to materialize their distribution along the transect via a georeferenced image, as shown in figure 5.

Table 2: Description of study sites

Sites	Altitude	Geographical coordinates	Habitat types observed	Surveyed area	Human actions
Kaziramihunda1	2128 m	S 03°12.771' E 029°32.810'	Plant formation in *Macaranga kilimandscharica* and	40 m x 30 m	-
Rwahirirwa	2192 m	S 03°12.798' E 029°32.698'	Plant formation in *Syzygium parvifolium*	40 m x 30 m	Cutting dead wood
Kaziramihunda 2	2169 m	S 03°12.820' E 029°32.653'	Formation plant *Prunus africana*	à 40 m x 40 m	-
Ndubura 1	2216 m	S 03°12.908' E 029°32.414'	Herbaceous scrub	30 m x 30 m	-
Ndubura 2	2260 m	S 03°12.913' E 029°32.380'	*Chrysophyllum gorungosanum*, *Macaranga* plant formation	40 m x 30 m	-
Ndubura 3	2281 m	S 03°12.887' E 029°32.312'	Plant formation in *Symphonia globulifera* and *Chrysophyllum gorungosanum*	30 m x 30 m	-

Fig. 5: Georeferenced image illustrating the transect traced in the Teza sector of KNP (Author)

III.2.2. Collecting specimens

Fieldwork took place from April 29, 2013 to April 28, 2013.
June 2013 and from December 16 to 30, 2013.

The site was subdivided into 10 m quadrants using stakes and ropes. Each quadrant was in turn subdivided into 100 1 m x 1 m squares of the same dimensions to facilitate the floristic inventory. The inventory was carried out systematically, tile by tile and stratum by stratum, starting with the uppermost stratum.

Two specimens of each plant species were collected and stored in newspaper, along with their vernacular names and coverings. At least two photos were taken of each specimen. Specimens of unrecognized species were marked with a number. In addition, species abundance was carefully observed at each site, enabling us to estimate the cover of each species by assigning it an abundance-dominance coefficient.

Table 3: Dominance abundance coefficient according to Braun-Branquet (1932) cited by Lewalle (1972)

Average coverage	Abondance	Weighting coefficient
Species present	+	0,1
0-5%	1	2,5
5-25%	2	15
25-50%	3	37,5
50-75%	4	62,5
75-100%	5	87,5

III.2.3. Dendrometric measurements

After harvesting was completed, circumference measurements of standing trees were taken using a tape measure from 1.30 m above the ground for all trees with a circumference greater than or equal to 15 cm (Malaise, 1984).

The heights of trees in open areas were determined using the Suunto Dendrometer. For trees in unclear areas, a previously measured pole was used to approximate their heights.

III.2.4. Litter quantification

Two tools were used to quantify the litter present at the various sites sampled: the first consisted of two solid sticks attached to a small board for collecting the litter (fig. 6); the second, which measures the size of the elements making up the litter, is a floor with grids, the first of which, taken as a unit, measures 2 cm on a side. The litter elements were collected by pushing the collection tool into the litter 30 times at each site, at random. The size of each item was measured by applying it to the floor, with litter items occupying levels 1 and 2 of the floor being considered small and level 1 as fine litter. This method was developed by Nzigidahera (2012).

a: Litter collection tool consisting of two solid sticks attached to a small board

b: Floor with grids for measuring litter size

Fig. 6: Litter analysis tools. Photos: D. Nzoyisaba, 2013

For each site, the number and size of all items collected were recorded. All observations, including human actions and signs of wildlife passage, were recorded in our field log, and a photo showing the physiognomic aspect of each site was taken. We finished by recording the geographical coordinates of each site using a GARMIN GPS (3m accuracy).

III.3. ANALYSIS METHODS

III.3.1. Identification of specimens

Specimens were identified using the four volumes of Troupin (1978-1988), *"Flore du Rwanda, Les Spermatophytes"*, *Manuel de Botanique forestière*, Tome 1 et 2 (Letouzey, 1983), and *Botanique systématique des plantes à* fleurs

(Rodolphe et *al.*, 2012). The herbarium of the University of Burundi and that of OBPE (formerly INECN) were consulted. To update our nomenclature, we referred to Lubrun and Stork's system (Enumeration of flowering plants of tropical Africa), online at http://www.ville- ge.ch/musinfo/bd/cjb/africa/ visited 05/02/ 2015.

III.3.2. Analysis of biological forms

Biological form shows a species' degree of resistance to adverse climatic conditions. Each ecosystem is characterized by the dominance of a specific biological form. The analysis of biological forms has been based mainly on the work of Lewale (1972), Habiyaremye (1995), Hakizimana (2012) and Masharabu (2012).

According to Raunkiaer (1934), cited by Hakizimana (2012), Burundi species are grouped into the following 5 categories:

- Phanerophytes (P): plants whose stem apparatus bears more than one 40 cm from the ground of protected evergreen buds;
- Chamephytes (Ch): plants with a dwarf vegetative apparatus less than 40 cm high, with persistent buds protected by plant debris;
- Hemicryptophytes (H): individuals characterized by an aerial vegetative apparatus that dries out completely during the off-season and whose persistent buds develop at the crown;
- Geophytes (Geo): plants with deciduous stems, buds and young shoots in the ground (bulbs, rhizomes and tubers);
- Therophytes (Th): annual plants that spend the off-season as seeds.

The spectrum of biological forms provides valuable insights into community structure, physiognomy and adaptive strategies (Gillet, 2000, cited by Masharabu, 2012). The gross biological spectrum expresses how the plant species making up a vegetation are distributed between biological forms. It is denoted by the following formula:

$$S.B_i = \frac{F_i}{N} \times 100$$

Where

$S.B_i$: raw spectrum of a given biological form;

F_i : number of times species of a given biological form are in all statements;

N: refers to total species.

The weighted biological spectrum uses Braun-Blanquet abundance-dominance coefficients as follows: species are assigned a coefficient corresponding to their abundance in each survey. The number of occurrences of each abundance-dominance coefficient is then determined for each biological trait and for all surveys, and multiplied by the corresponding weighting coefficient. For example, the total number of "+" values will be multiplied by 0.1, while the total number of "1" values will be multiplied by 2.5, and so on up to the total number of "5" values, which will be multiplied by 87.5. The figures obtained are then added together for each biological trait. The grand total of all the readings is brought back to 100 and the sums for each category are adjusted to the percentage (Masharabu, 2012).

The weighted biological spectrum is given by the following expression:

$$S.P_i = \frac{n_i}{N} \times 100$$

Where

$S.P_i$: weighted spectrum;

n_i : the share of a biological form;

N: the sum of the shares of all biological forms.

III.3.3. Analysis of phytogeographical features

Lebrun defines the "base element" as the set of species characteristic of a region that do not (or hardly) transgress its boundaries (Troupin, 1966), quoted by Hakizimana (2012). This element is the most perfect expression of the phytogeographical individuality of a given region; it may include sub-elements limited to part of that region. Determining the phytogeographical elements to which the identified species belong provided information on their distribution area.

We referred mainly to the works of Lewale (1972), Habiyaremye (1995), Hakizimana (2012) and Masharabu (2012). The following phytogeographical elements have been selected:

Widespread species:

- Pantropical (Pan): species found in all tropical regions of the world (Asia, Africa, America);
- Palaeotropical (Pal): species found in tropical Asia, America, Madagascar and Australia;
- Pluriregional (Plur): species that are widely distributed across the globe, sometimes even spanning different floral empires;
- Cosmopolitan (Cos): species distributed in tropical and temperate regions.

Linking species

These are species distributed in two or three regions and sharing a geographical boundary.
- L.SZ-G: Species found in the Guinean, Sudanian and Zambezi regions.

Mountain species (Mo): These are species whose range extends over most of the mountains of Africa.

SudanoZambezian (SZ) species: These are species distributed in Sudano-Zambezian regions, we have:

- SZ: Sudano-Zambézian or omni-Sudano-Zambézian species;
- SZ (O): Sudano-Zambézian species with eastern dominance;
- SZ (Z): Sudano-Zambézian species with Zambézian dominance;
- SZ(OZ): Sudano-Zambézian species with Eastern and Zambézian dominance;
- SZ (EO): Sudano-Zambézian species with Ethiopian and Oriental dominance;
- SZ(EOZ): Sudano-Zambézian species with Eastern Ethiopian and Zambézian dominance.

Endemic species (End): distribution limited to Africa
central, as defined by Lewalle (1972).

III.3.4. Face analysis

We studied stratification based on tree height measurements. Height measurements were taken on all trees whose circumference at breast height (1.30 m) was greater than or equal to 15 cm. The distinction between the different strata was based on the OBPE (ex INECN) LEM sheet as designed by Habiyaremye (2012). The following strata were selected:

- the arborescent stratum with very large trees (**A-TGA**): 30 m < 50 m
- tree stratum with large trees (**A-GA**): 20 m < 30 m
- the arborescent stratum made up of small to medium-sized trees (**A-AM**): 7 m < 20 m
- shrub layer (**aB**): 2 m < 7 m
- sub-shrubby and/or herbaceous stratum (**SsAH**):< 2 m

By careful observation in the field, overlap (the fraction expressed as a % of the surface area occupied by the ground projection of the twigs and foliage of each stratum) was estimated and attributed to each stratum (Manirakiza, 2013).

III.3.5. Basal area

Basal area now plays an increasingly important role in forest management. It is used in stand diagnosis, silvicultural itinerary monitoring, and development and management planning. It is also used to measure the intensity of competition between trees in a stand (Cordonnier, 2014).

We calculated basal area using the following formula:

$$g_i = \frac{c_i^2}{4\pi}$$

Where g_i : basal area of a tree i at 1.30 m height in m /ha^2

c_i : circumference of a tree bole at 1.30 m

The total basal area was calculated by classes of 10 cm circumference intervals and by species found at each site, using the following formula:

$$\text{where } Gi: \quad Gi = Ni x \frac{c^2}{4\pi} \quad \text{basal area in a class (m²/ha).}$$

C: Average circumference in a class

Total basal area is obtained by summing all basal areas by class or species. The basal area for a given site is equal to the sum of all basal areas of individuals of species present on the surveyed area, per hectare.

The different circumferences measured were grouped into successive classes. A graph showing the distribution of the number of stems according to circumference classes was drawn up for each site.

III.3.6. Stand density

For wood density, the number of stems per hectare was estimated for each site and for the entire study area. The various stems were divided into circumference classes ranging from 15 cm to 10 cm intervals. After plotting the distribution of stems according to their circumference classes, a logarithmic trend curve was added:

$$y = a \ln x + b$$

Where y is the number of stems; x is the circumference class i, a and b are any real numbers depending on the stem distribution, $a < 0$

Indeed, as the circumference increases, the curve takes on a logarithmic shape, which could be an indication of stability according to Troupin (1966) and Bouxin (1973) cited by Bukobero (1998). To find out at what level the logarithmic trend is followed by the observed distribution, we calculated the coefficient of determination R^2 between the two distributions.

To properly interpret the results found for basal area, litterfall and wood density, Pearson's sampling correlation coefficient was calculated.

III.3.7. Habitat dynamics

To be able to propose the probable evolution of our study sites, the species making up the main strata inventoried were highlighted with their respective heights and cover. A formula characterizing the current vegetation of each site was established. From the current state thus characterized, an attempt was made to predict the evolution of the plant formation in place. This formula is of the

ABSHG type, where

A: Tree stratum B: Shrub stratum

S: Shrub stratum H: Herbaceous stratum

G: Grass stratum

Each stratum was assigned an index corresponding to its coverage according to the scale used by Troupin (1966), which gives the percentage of the surface occupied by each stratum (Table 4).

Finally, for each of the 6 sites, a LEM form was drawn up to record observations on habitat dynamics. These sheets can be used as a reference for monitoring the dynamics of the habitats surveyed.

Taleau 4: Level of overlap (Troupin, 1966)

Index	Recovery (%)
10	91-100
9	81-90
8	71-80
7	61-70
6	51-60
5	41-50
4	31-40
3	21-30
2	11-20
1	5-10
+	<5

III.3.8. Results analysis

• Floristic similarity between sites

The comparison of the flora of different sites, two by two, was made on the basis of Sorensen's similarity index (K) (1984) cited by Hakizimana (2012). The advantage of this index is that it is adapted to qualitative, presence/absence binary data and gives more weight to species common to the plant formations being compared by multiplying their number by two. It does not take into account species absent from the two stands compared but present in other nearby stands. This index is expressed in % and its values vary from 0 to 100, respectively when the two plant formations compared have few or no species in common, or when the two formations have many species in common, and are therefore floristically identical. These values are obtained using the following formula.

$$K = \frac{2a}{2a+b+c} \times 100$$

Where

a: number of species common to the two sites compared
b and c: numbers of species absent in one of the two sites compared but present in the other

MVSP (MultiVariate Statistical Package) software (Kovach, 2003 cited by Bangirinama, 2010) was used to detect floristic dissimilarities between the 6 sites surveyed. We opted for the UPGMA (Unweighted Pair Group Method with Arithmetic mean) method, which has the advantage of giving the same weight ("unweighted") to the different objects compared (surveys or hills) and is the most widely used (Sokal & Michener, 1958; Senterre, 2005; Bangirinama, 2010, Hakizimana, 2012). Thanks to the Cluster Analysis technique, the matrix of data subjected to the method is reduced at each stage, and each object is given the same weight.

The highest similarity then enables the method technique to select the next cluster to be formed. A dendrogram showing the layout of the 6 sites surveyed was obtained.

- **Pearson correlation coefficient**

Pearson's correlation coefficient (r) was used in this work to measure the degree of linearity between two sets of data whenever necessary. The value of this coefficient varies between -1.0 and 1.0 inclusive. The further the coefficient value is from zero, the better the correlation. For example, if r = +1, there is a perfect positive correlation; if r = -1, there is a perfect negative correlation; and if r = 0, there is no correlation at all.

$$r = \frac{\sum (x-\bar{x})(y-\bar{y})}{\sqrt{\sum (x-\bar{x})^2 \sum (y-\bar{y})^2}}$$

where $\bar{x}$ and $\bar{y}$ are the sample means

- **Specific quotient**

The degree of maturity and stability of the flora at the various sites sampled was estimated on the basis of the value of the specific quotient
(Q) (Evrard, 1968 in Sonké, 1998 cited by Hakizimana *et al*, 2011) which is obtained by the following formula:

$Q = \dfrac{S}{Ge}$ where S is the number of species identified at each site and Ge the number of genres.

IV. RESULTS AND DISCUSSION

IV.1. PRESENTATION AND INTERPRETATION OF RESULTS

IV.1.1. Floristic composition

IV.1.1.1. General considerations

Throughout the study area, 166 species in 132 genera and 73 families were inventoried, of which 4 were determined down to the genus level (Table 5). This vegetation is largely dominated by the Dicotyledons (magnoliopsida) class, with 55 families (75.34%), 108 genera (81.82%) and 129 species (77.71%). The Monocotyledons (liliopsida) class is represented by 9 families (12.32%), 15 genera (11.36%) and 18 species (10.84%). The

Pteridophytes comprise 8 families (10.95%), 8 genera (6.11%) and 18 species (10.40%). The Bryophytes class is very poorly represented, with 1 species (1.37%). It should be noted that 3 species were determined down to family level. Some of the species collected did not have characteristics that facilitated identification, such as flowers or fruits, which is why not all the species inventoried could be identified. A detailed list of all species identified, family by family, can be found in Appendix 1.

The most represented families are the Rubiaceae with 14 species (8.43%), the Asteraceae with 8 species (4.82%), the Acanthaceae with

7 species (4.22%), Aspleniaceae with 8 species (4.82%), Orchidaceae and Euphorbiaceae with 7 species each (4.22%) (fig. 7). Urticaceae and Verbenaceae are represented by 6 species each (3.61); the other families are poorly represented, 43 of which are monospecific.

Table 5: Floristic composition by family in the study area. Percentages were calculated on the basis of the grand total.

Families	Genres	%	Species	%
1. DICOTYLEDONES (MAGNOLIOPSIDA)				
Acanthaceae	7	5,30	7	4,22
Alangiaceae	1	0,76	1	0,60
Amaranthaceae	1	0,76	1	0,60
Annonaceae	1	0,76	1	0,60
Apocynaceae	2	1,52	2	1,20
Araliaceae	2	1,52	3	1,81
Asclepiadaceae	3	2,27	3	1,81
Aspidiaceae	1	0,76	1	0,60
Asteraceae	7	5,30	8	4,82
Balsaminaceae	1	0,76	4	2,41
Basellaceae	1	0,76	1	0,60
Begoniaceae	1	0,76	1	0,60
Campanulaceae	1	0,76	1	0,60
Caryophyllaceae	1	0,76	1	0,60
Celastraceae	1	0,76	1	0,60
Chrysobalanaceae	2	1,52	2	1,20
Clusiaceae	2	1,52	2	1,20
Connaraceae	1	0,76	1	0,60
Convolvulaceae	1	0,76	1	0,60
Crassulaceae	1	0,76	1	0,60
Cucurbitaceae	3	2,27	3	1,81
Euphorbiaceae	7	5,30	7	4,22
Fabaceae	5	3,79	5	3,01
Flacourtiaceae	1	0,76	1	0,60
Geraniaceae	1	0,76	1	0,60
Hippocrateaceae	1	0,76	1	0,60
Lamiaceae	3	2,27	3	1,81
Malvaceae	1	0,76	1	0,60
Meliaceae	1	0,76	1	0,60
Melianthaceae	1	0,76	1	0,60
Menispermaceae	1	0,76	1	0,60
Monimiaceae	1	0,76	1	0,60
Moraceae	1	0,76	1	0,60
Myrsinaceae	2	1,52	3	1,81
Myrtaceae	1	0,76	1	0,60
Olacaceae	1	0,76	1	0,60
Oleaceae	1	0,76	3	1,81
Passifloraceae	1	0,76	1	0,60
Phytolaccaceae	1	0,76	1	0,60
Piperaceae	2	1,52	4	2,41
Polygonaceae	1	0,76	1	0,60
Ranunculaceae	2	1,52	2	1,20
Rhamnaceae	1	0,76	1	0,60
Rosaceae	2	1,52	3	1,81
Rubiaceae	11	8,33	14	8,43
Rutaceae	1	0,76	1	0,60
Sapindaceae	1	0,76	2	1,20
Sapotaceae	3	2,27	3	1,81
Solanaceae	2	1,52	2	1,20
Sterculiaceae	1	0,76	1	0,60
Theaceae	1	0,76	1	0,60
Tiliaceae	1	0,76	1	0,60
Urticaceae	5	3,79	6	3,61
Verbenaceae	1	0,76	6	3,61
Vitaceae	1	0,76	1	0,60

Family				
Total	**108**	**81,82**	**129**	**77,71**
2.MONOCOTYLEDONES (LILIOPSIDA)				
Anthericaceae	1	0,76	1	0,60
Araceae	2	1,52	2	1,20
Commelinaceae	1	0,76	1	0,60
Cyperaceae	1	0,76	2	1,20
Dracaenaceae	1	0,76	1	0,60
Musaceae	1	0,76	1	0,60
Orchidaceae	5	3,79	7	4,22
Poaceae	2	1,52	2	1,20
Smilacaceae	1	0,76	1	0,60
Total	**15**	**11,36**	**18**	**10,84**
3. PTERIDOPHYTES				
Aspleniaceae	1	0,76	8	4,82
Cyatheaceae	1	0,76	2	1,20
Dennstaedtiaceae	1	0,76	1	0,60
Dryopteridaceae	1	0,76	3	1,81
Oleandraceae	1	0,76	1	0,60
Polypodiaceae	1	0,76	1	0,60
Pteridaceae	1	0,76	1	0,60
Vittariaceae	1	0,76	1	0,60
Total	**8**	**6,11**	**18**	**10,84**
4. BRYOPHYTES				
Hymenophyllaceae	1	0,76	1	0,60
Total	**1**	**0,76**	**1**	**0,60**
Grand total	**132**	**100,00**	**166**	**100,00**

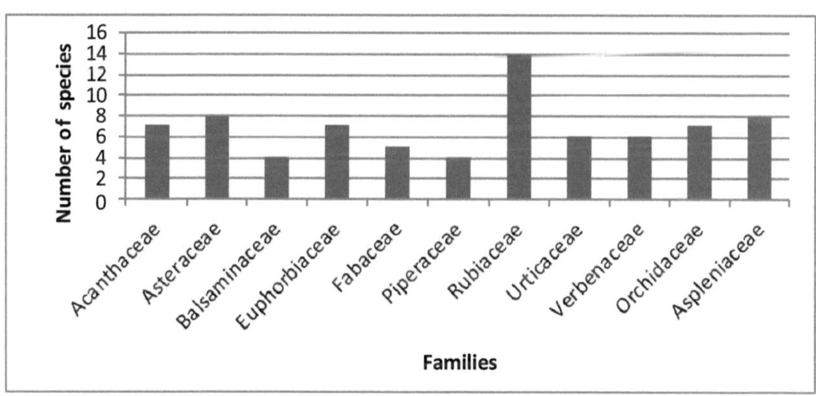

Fig. 7: Most represented families in Kibira National Park (Secteur Teza)

IV.1.1.2. Floristic characteristics of study sites

- **Floristic richness of various sites**

The various sites surveyed show differences in terms of floristic composition. Table 6 shows the floristic richness of all the sites surveyed.

- *Kaziramihunda 1 site*

Seventy-five species (45.18% of the flora of the sites sampled) were counted in 66 genera and 53 families. Dicotyledons came first with 57 species, i.e. 76%, followed by Monocotyledons with 9 species, i.e. 12%, Pteridophytes came third with 8 species, i.e. 10.67%, and finally one Bryophyte species, i.e. 1.33%.

The Rubiaceae family is the most represented with 9 species (12%), followed by the Piperaceae with 4 species (5.33%).

Balsaminaceae and Araliaceae are also represented, with 3 species (4%) each. The others are poorly represented. This is the most phytodiverse site after Kaziramihunda 2.

- *Rwahirirwa site*

With 67 species, Rwahirirwa accounts for 40.36% of the flora of the sites sampled, divided into 60 genera and 42 families. Dicotyledons dominate with 51 species (76.12%), followed by Monocotyledons with 9 species (13.43%) and Pteridophytes with 7 species (10.45%). No Bryophyte species were found on this site. Rubiaceae are better represented with 8 species (11.94%), Acanthaceae, Aspleniaceae, Euphorbiaceae and Myrsinaceae are also represented with 3 species (4.48%) each.

- *Kaziramihunda 2 site*

With 104 species (62.65% of the flora of the sites surveyed) divided into 91 genera and 55 families, Kaziramihunda 2 has many species compared with the other sites. Dicotyledons take the lead with
76 species (73.08%), followed by Monocotyledons with 15 species (14.42%), Pteridophytes with 12 species (11.54%) and Bryophytes with 1 species (0.96%).

Rubiaceae dominate with 10 species (9.62%), Urticaceae and Orchidaceae with 6 species (5.77%) each, Euphorbiaceae and Fabaceae with 5 species (4.81%) each, Aspleniaceae and Piperaceaee with 4 species (3.85%) each.

- *Ndubura 1 site*

We counted 55 species, or 33.13% of the flora of the entire study area, divided into 51 genera and 37 families. Dicotyledons were in first place with 46 species (83.64%), followed by Monocotyledons with 5 species (9.09%) and Pteridophytes with 4 species (7.27%). Acanthaceae and Euphorbiaceae take the lead with 4 species (7.27%) each, Rubiaceae, Urticaceae and Asteraceae come second with 3 species (5.45%) each. The others are represented by less than 3 species.

- *Ndubura 2 site*

Ndubura 2 has fewer species than the other sites. There were 50 species (30.12%) in 47 genera and 34 families. Dicotyledons dominate with 43 species (86%), followed by Pteridophytes with 5 species (10%) and Monocotyledons with 2 species (4%). Rubiaceae are better represented with 6 species (12%), Asteraceae with 4 species (8%) and Euphorbiaceae with 3 species (6%). The other families are less well represented, 25 of which are monospecific.

- *Ndubura 3 site*

There were 55 species (33.13%) divided into 52 genera and 34 families. Dicotyledons dominate with 46 species (83.64%), followed by Pteridophytes with 5 species (9.09%) and Monocotyledons with 4 species (7.27%). Rubiaceae dominate with 5 species (9%), Acanthaceae come second with 4 species (7.27%), Asteraceae, Aspleniaceae, Euphorbiaceae and Verbenaceae are represented by 3 species (5.45%) each.

Table 6: Floristic richness of surveyed sites

Families	Kaziramihunda 1	Rwahirirwa	Kaziramihunda 2	Ndubura 1	Ndubura 2	Ndubura 3
Broadleaf	57	51	76	46	43	46
Acanthaceae	0	3	3	4	0	4
Amaranthaceae	1	1	1	2	1	1
Annonaceae	1	1	0	0	0	0
Apocynaceae	2	2	2	0	2	2
Araliaceae	3	2	3	0	1	2
Asclepiadaceae	2	2	2	1	1	2
Aspidiaceae	0	0	1	0	0	0
Asteraceae	1	2	3	3	4	3
Balsaminaceae	3	2	2	2	0	1
Basellaceae	0	0	0	1	0	0
Begoniaceae	1	1	1	1	1	0
Campanulaceae	0	1	1	0	0	0
Alangiaceae	1	0	1	0	0	1
Caryophyllaceae	0	0	0	1	0	0
Celastraceae	1	1	1	0	1	0
Chrysobalanaceae	1	0	0	0	1	2
Clusiaceae	1	1	1	0	1	1
Connaraceae	1	1	1	0	1	1
Convolvulaceae	1	1	0	0	0	0
Crassulaceae	1	0	1	0	0	0
Cucurbitaceae	2	2	2	2	1	0
Euphorbiaceae	1	3	5	4	3	3
Fabaceae	0	1	5	1	1	1
Flacourtiaceae	1	0	1	0	0	1
Geraniaceae	0	0	0	1	0	0
Hippocrateaceae	0	0	0	0	1	1
Lamiaceae	1	2	1	0	1	1
Malvaceae	0	0	0	1	0	0
Meliaceae	0	0	1	0	0	0
Melianthaceae	1	1	1	0	0	0
Menispermaceae	1	1	1	1	0	0
Monimiaceae	1	1	1	0	1	1
Moraceae	1	1	1	0	1	1
Myrsinaceae	2	3	3	1	1	1
Myrtaceae	1	1	1	0	1	1
Olacaceae	1	0	0	0	1	1
Oleaceae	1	0	2	0	2	0
Passifloraceae	0	0	0	1	1	0
Phytolaccaceae	0	0	0	1	0	0
Piperaceae	4	1	4	1	1	1
Polygonaceae	0	0	0	1	0	0
Ranunculaceae	0	0	0	2	0	0
Rhamnaceae	0	0	1	0	0	0
Rosaceae	1	1	1	2	0	0
Rubiaceae	9	8	10	3	6	5
Rutaceae	1	0	0	0	1	0
Sapindaceae	1	1	1	0	1	1
Sapotaceae	1	0	1	0	2	2
Solanaceae	0	0	0	2	0	0

Sterculiaceae	0	0	0	1	0	0
Theaceae	1	1	1	0	0	0
Tiliaceae	0	1	0	1	0	0
Urticaceae	1	1	6	3	1	2
Verbenaceae	2	0	1	1	2	3
Vitaceae	1	0	1	1	0	0
Monocotyledons	**9**	**9**	**15**	**5**	**2**	**4**
Anthericaceae	1	0	0	0	0	0
Araceae	2	1	2	1	1	2
Commelinaceae	0	0	1	0	0	0
Cyperaceae	2	2	2	1	0	0
Dracaenaceae	1	1	1	1	1	1
Musaceae	0	0	1	1	0	0
Orchidaceae	0	2	6	0	0	0
Poaceae	2	2	1	1	0	0
Smilacaceae	1	1	1	0	0	1
Pteridophytes	**8**	**7**	**12**	**4**	**5**	**5**
Aspleniaceae	2	3	5	1	2	3
Cyatheaceae	1	1	1	1	1	0
Dennstaedtiaceae	0	0	0	1	0	0
Dryopteridaceae	2	1	3	1	1	1
Oleandraceae	0	1	1	0	1	0
Polypodiaceae	1	1	1	0	0	1
Pteridaceae	1	0	1	0	0	0
Vittariaceae	1	0	0	0	0	0
Bryophytes	**1**	**0**	**1**	**0**	**0**	**0**
Hymenophyllaceae	1	0	1	0	0	0
Total	**75**	**67**	**104**	**55**	**50**	**55**
%	**45,1**	**40,3**	**62,6**	**33,13**	**30,1**	**33,13**

- **Floristic similarity between sites**

The flora of different sites was compared using Sorensen's index of similarity (Table 7). Generally speaking, the values found show that most sites have many common species. In fact, the K values found between certain sites show that more than 50% of species are common. This is the case for Kaziramihunda 1 and Rwahirirwa (K = 60.7), Kaziramihunda 1 and Kaziramihunda 2 (K = 62.3), Kaziramihunda 2 and Rwahirirwa (K = *60.7*), Kaziramihunda 1 and Kaziramihunda 2 (K = 62.3) and Kaziramihunda 2 and Rwahirirwa (K = *62.3)*. (*K=54*.7), Ndubura 2 and Ndubura 3 (K=56.4), Kaziramihunda 2 and Ndubura 3 (K=50.9). Sites with few common species are Ndubura 1 and Ndubura 3 (*K=22*.4), Ndubura 1 and Ndubura 2 (K= 24.5), Ndubura 1 and Kaziramihunda 1 (K= 28.4).

Table 7: Similarity indices for different sites

Stes	Kaziramihunda 1	Rwahirirwa	Kaziramihunda 2	Ndubura 1	Ndubura 2	Ndubura 3
Kaziramihunda 1		60,7	62,3	28,4	48,4	49,3
Rwahirirwa			54,7	32,5	47,9	36,2
Kaziramihunda 2				32,29	45,2	50,9
Ndubura 1					24,5	22,4
Ndubura 2						56,4

A comparison of floristic composition using the dendrogram (Fig.8) reveals a clear dissimilarity between different sampled sites. Five groupings are identifiable, showing that most sites retain their identity. The similarity between the flora of Kaziramihunda 1 (secondary forest) and that of Kaziramihunda 2 (primary forest) stems from the fact that the two sites are located in comparable ecological conditions. In fact, both sites belong to the middle horizon (Lebrun, 1935 cited by Lewalle, 1972) and are located near valleys. The floristic dissimilarities observed between Ndubura 1 and the other sites would seem to be linked to disturbance.

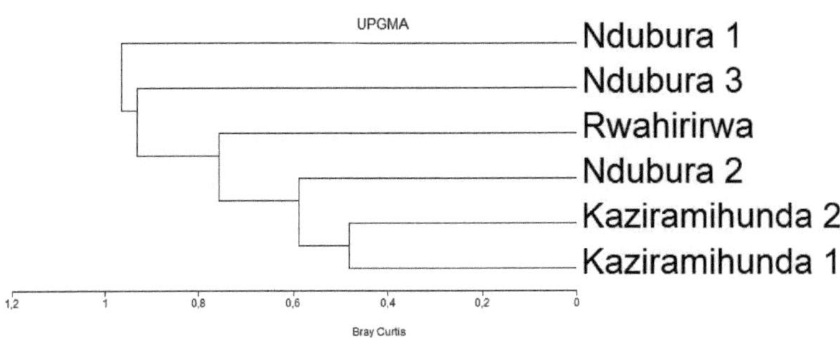

Fig. 8: Dendrogram of surveyed sites in KNP (Teza sector)

- **Site-specific quotas**

Table 8 shows the specific quotients (Q) for all the sites surveyed. The Q values found for all sites are low, reflecting the maturity of their flora.

Table 8: Specific quotients for prosperous sites

Sites	Number of species	Number of types	Q
Kaziramihunda 1	75	66	1.14
Rwahirirwa	67	60	1.12
Kaziramihunda 2	104	91	1.14
Ndubura 1	55	51	1.08
Ndubura 2	50	47	1.06
Ndubura 3	55	52	1.06

IV.1.2. Biological forms

- **General appearance**

Of a total of 166 species determined, 127 were analyzed for biological forms. After analysis, it was found that 81 species are phanerophytes, i.e. 63.8% of the gross spectrum, 19 species are chamephytes, i.e. 15% of the gross spectrum, the other biological forms are poorly represented, namely geophytes (9.4%), therophytes (6%), hemicryptophytes (4%) and epiphytes (2%).

However, when analyzed by weighted spectra, the order of these biological forms changes, with phanerophytes in first place at 93.3%, followed by therophytes at 2.6%, followed by chamephytes (2.2%), geophytes (1.4%), hemicryptophytes (0.3%) and epiphytes (0.05%) (fig.9). Given the importance of phanerophytes, this study area retains its forest character. The non-negligible proportion of champhytes in the gross spectrum may be linked to the stress tolerance strategy. In our study area, stresses are largely of a luminous nature in the forest undergrowth. Geophytes are the biological form best adapted to environments that are both stressed and disturbed. In the various sites surveyed, geophytes are dominated by species of the Aspleniaceae family, such as *Asplenium dregeanum, Asplenium elliotii, Asplenium aethiopicum, Asplenium mannii* and *Asplenium friesiorum*.

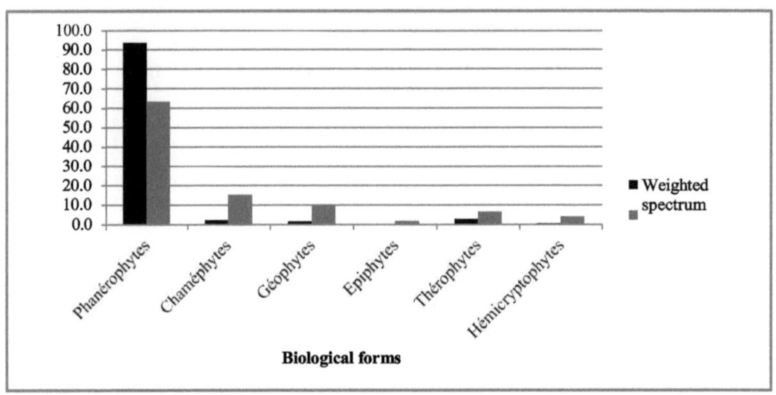

Fig. 9: Crude and weighted spectrum of biological forms of species in the area study

- **Biological forms at different sites**

- *Kaziramihunda 1 site*

Of the 62 species for which biological forms were analyzed, 43 were phanerophytes, representing 68.9% of the raw spectrum, followed by chamephytes with 16.4%, geophytes with 8.2% and therophytes and hemicryptophytes with 3.3% each. Considering the weighted spectra, phanerophytes lead with 91.22%, chaméphytes follow with 3.22%, therophytes and hemicryptophytes are at the same level with 1.95% and finally geophytes with 1.63% (fig.10). The high proportions of phanerophytes bear witness to the forested nature of the site. The significant proportions of champhytes in the weighted spectrum are due to the presence of light-tolerant species such as *Impatiens stuhlmannii*, *Impatiens purpureo-violacea*, *Kalanchoe crenata*, *Momordica foetida* and *Cyperus pseudoleptocladus*.

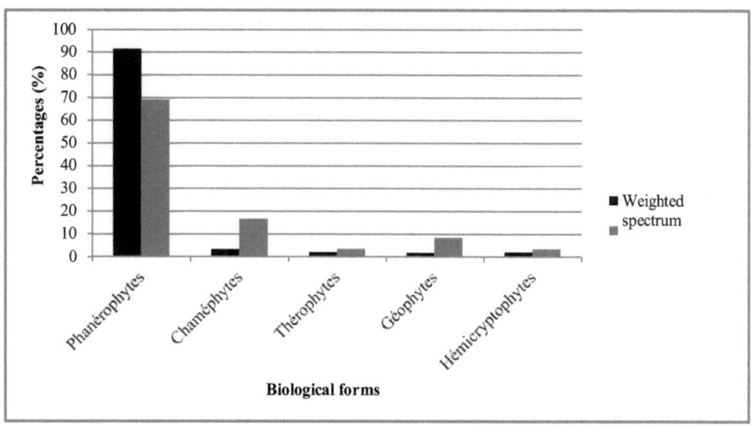

Fig. 10: Crude and weighted spectrum of species biological forms at the Kaziramihunda 1 site

- *Rwahirirwa site*

The proportions of biological forms (Fig.11) show that phanerophytes (66% of the raw spectrum) far outstrip the other biological forms, which are chamephytes (17%), geophytes (10%), therophytes (3%), hemicryptophytes (2%) and epiphytes (2%). Analysis of the weighted spectra still places phanerophytes in first place with 76.8%, followed by chamephytes (10.8%), therophytes (10.3%) and geophytes (2%). Hemicryptophytes and epiphytes account for a small proportion at 0.1%. The importance of phanerophytes in relation to other biological forms is indicative of the site's forested character. The high proportion of chamephytes may be linked to the adaptation strategy to light stress. The importance of therophytes in the weighted spectrum and of geophytes in the gross spectrum could provide information on the disturbance of the site. In fact, the site is located near the Bukeye-Rugazi track, and cuttings of dead wood have been observed.

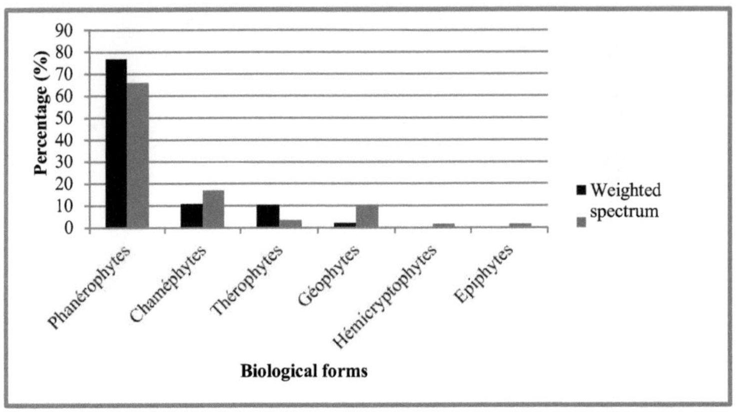

Fig. 11: Crude and weighted spectrum of species biological forms at the Rwahirirwa site

- *Kaziramihunda 2 site*

At this site, analysis of biological forms reveals the dominance of phanerophytes (63.4% of the gross spectrum), with chamephytes in second place at 15.9%. Geophytes (8.5%), therophytes (6%), epiphytes (3.7%) and hemicryptophytes (2.4%) are also present, but at low levels. Compared in terms of weighted spectra, phanerophytes come first with 96%. They are followed by geophytes (2%), chamephytes (1%), hemicryptophytes (0.25%), epiphytes (0.25%) and therophytes (0.3%) (fig.12). The high proportion of phanerophytes indicates that the site is forested. The importance of the Aspleniaceae family at Kaziramihunda 2 would explain the high rate of geophytes.

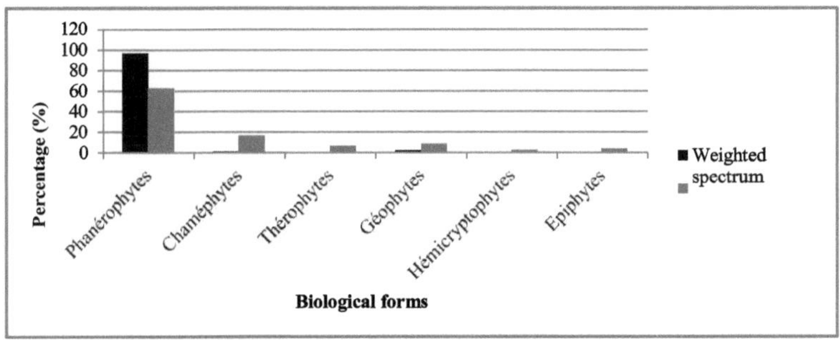

Fig. 12: Crude and weighted spectrum of species biological forms at the Kaziramihunda 2 site

- *Ndubura 1 site*

In terms of the raw spectrum, phanerophytes come first with 70%, followed by chaméphytes with 13%, therophytes and geophytes with 6% and finally hemicryptophytes with 5%. Considering the weighted spectrum, phanerophytes predominate (94.9% of the weighted spectrum). Therophytes (4.2%) come second. The other biological forms present a low rate and are chamephytes (0.5%), geophytes (0.2%) and hemicryptophytes (0.2%) (Fig.13). The high proportion of phanerophytes highlights the forested nature of the site. The importance of therophytes indicates the existence of ruderal and post-cultural species, such as *Impatiens burtonii*, *Solanum nigrum* and *Pilea bambuseti*, providing information on previous disturbances to the site.

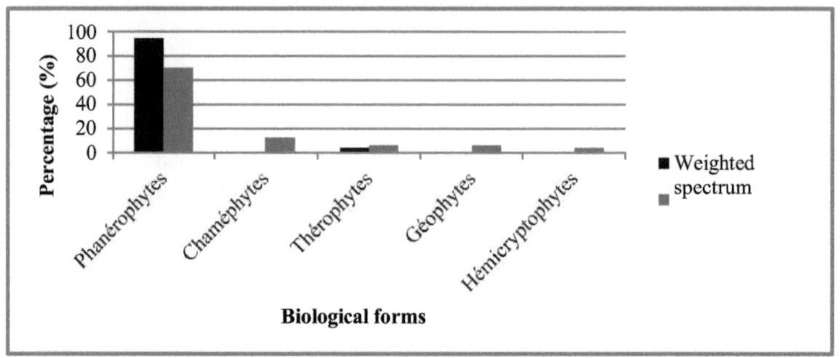

Fig. 13: Crude and weighted spectrum of species biological forms at the Ndubura 1

- *Ndubura 2 site*

Compared in terms of the number of individuals making up the biological forms, phanerophytes dominate, at 84%. They are followed by chamephytes at 11%, then geophytes at 5%. When the weighted spectrum is analyzed, phanerophytes take the lead with 98.5%, while geophytes (1.3%) and chamephytes (0.2%) present low proportions (Fig.14). Other biological forms are absent from the site, reflecting the dominance of woody species.

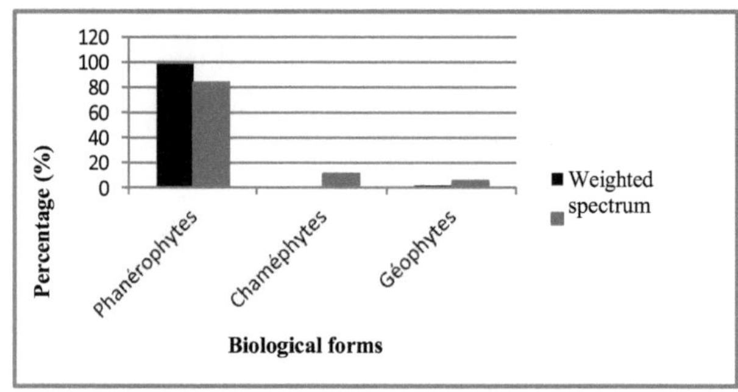

Fig. 14: Crude and weighted spectrum of species biological forms at the Ndubura 2

- *Ndubura 3 site*

Analysis of the biological forms (Fig. 15) shows the predominance of phanerophytes (82% of the raw spectrum). Geophytes come second with 9%. Other biological forms such as champhytes (7%) and therophytes (2%) are poorly represented. Looking at the weighted spectrum, the order remains the same: phanerophytes are in first place with 97.7%, geophytes come second with 2%, chamephytes and therophytes follow with 0.2% and 0.1% respectively. As at other sites, phanerophytes continue to dominate, reflecting the site's forested character. The importance of geophytes is linked to the presence of shade-tolerant species such as *Arisaema mildbraedii, Asplenium dregeanum, Asplenium elliotii* and *Loxogramme abyssinica.*

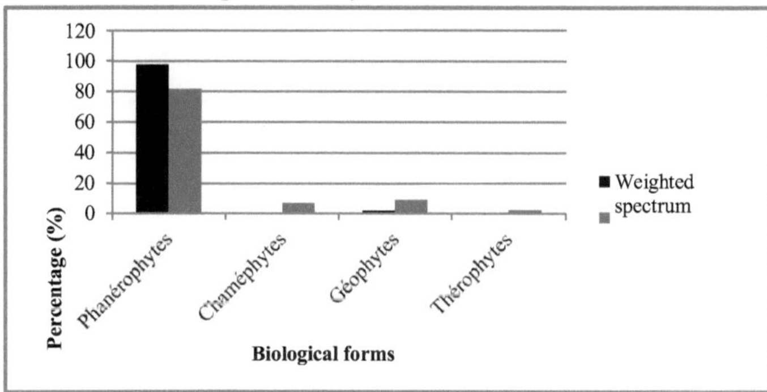

Fig. 15: Crude and weighted spectrum of species biological forms at the Ndubura 3

IV.1.3. Phytogeographical elements

• General appearance

Of the 166 species identified, the phytogeographical analysis was carried out on 126 species only, based on the documents available (Table 9). Montane species come first with 42.7%. Of these, 66% are strictly montane and 34% are East African montane species (Table 10). This shows that our study area retains a good number of species from the Afromontane domain in which it is located. In second place come Sudano-Zambézian species (17.7%), followed by Pluri-African species (10.5%). The others are poorly represented. The low proportions of wide-ranging species show that our study area is less disturbed.

Table 9: Geographical distribution of species harvested in Kibira (Teza)

Phytogeographical types	Number of species	%
Cosmopolites	2	1,6
Subcosmopolites	1	0,8
Paleotropics	5	3,2
Pantropicals	4	3,2
Wide-ranging species	**12**	**8,9**
Afro-Malagasy	8	6,5
Afro-Tropicales	6	4,8
Multi-African	13	10,5
More common African species	**27**	**21,8**
Mountain women	53	42,7
Sudano-Zambéziennes	23	17,7
Species with regional distribution	**76**	**60,5**
L-SZ-G	7	5,6
Linking species	**7**	**5,6**
Endemic	**2**	**1,6**
Guinean species	**2**	**1,6**
Total	126	100,0

Table 10: Selected groups of the montane base element in Teza forest (Mo: montane species; Mo(EA): East African montane species)

Groups	Number of groups	Raw spectrum (%)
Mo	35	66
Mo(EA)	18	34
Total	53	100

• Phytogeographical features of various sites

- Kaziramihunda 1 site

In the Kaziramihunda 1 site, montane species dominate with 49.2% of the total, placing this site in the Afromontane domain. They are followed by Sudano-Zambézian species (14.8%) and Pluri-African species (9.8%) (Table 11). The other phytogeographical elements are represented at a low level. The breakdown of species in the montane base element into typical montane groups (63.3%) and East African groups (36.7%) is shown in Table 12. Kaziramihunda 1 is therefore a mountain rainforest. The importance of sudano-zambézian species shows the probable disturbance of the site.

Table 11: Geographical distribution of species collected at the Kaziramihunda 1 site

Phytogeographical types	Number of species	%
Paleotropics	1	1,6
Pantropicals	2	3,3
Wide-ranging species	**3**	**4,9**
Afro-Malagasy	4	6,6
Afro-Tropicales	3	4,9
Multi-African	6	9,8
More common African species	**13**	**21,3**
Mountain women	30	49,2
Sudano-Zambéziennes	9	14,8
Species with regional distribution	**39**	**63,9**
L-SZ-G	3	4,9
Linking species	**3**	**4,9**
Endemic	**2**	**3,3**
Guinean species	**1**	**1,6**
Total	**61**	**100**

Table 12: Selected groups of the montane base element at the Kaziramihunda 1 site (Mo: montane species; Mo(EA): East African montane species)

Groups	Number of groups	Raw spectrum (%)
Mo	19	63,3
Mo(EA)	11	36,7
Total	30	100

- Rwahirirwa site

At this site, montane species predominate (57.6%), followed by Sudano-Zambézian species (10.2%), Pluri-African species (6.8%) and liaison species (6.8%). The others are poorly represented, as Table 13 shows. Among montane species, the omni-montane group (Mo) dominates with 58.8%, compared with 41.2% for the East African montane group Mo(EA) (Table 14). Such a phytogeographical distribution shows that Rwahirirwa belongs to the Afromontane domain.

Table 13: Geographical distribution of species collected at the Rwahirirwa site

Phytogeographical types	Number of species	%
Paleotropics	3	5,1
Pantropicals	2	3,4
Wide-ranging species	**5**	**8,5**
Afro-Malagasy	2	3,4
Afro-Tropicales	3	5,1
Multi-African	4	6,8
More widespread African species	**9**	**15,3**
Mountain women	34	57,6
Sudano-Zambéziennes	6	10,2
Species with regional distribution	**40**	**67,8**
L-SZ-G	4	6,8
Linking species	**4**	**6,8**
Endemic species	**1**	**1,7**
Total	**59**	**100**

Table 14: Selected groups of the montane base element at the Rwahirirwa site (Mo: montane species; Mo(EA): East African montane species)

Groups	Number of groups	Raw spectrum (%)
Mo	20	58,8
Mo(EA)	14	41,2
Total	**34**	**100**

- *Kaziramihunda 2 site*

Analysis of the geographical distribution of Kaziramihunda 2 species shows that montane species top the list with 47.4%, followed by Sudano-Zambézian (16.7%) and Pluri-African (14.1%) species. The others are poorly represented (Table 15). Montane species are subdivided into the omni-montane group (62.2%) and the East African montane group (37.8%) (Table 16). This phytogeographical distribution places Kaziramihunda in the afromontane domain.

Table 15: Geographical distribution of species collected at the Kaziramihunda 2 site

Phytogeographical types	Number of species	%
Subcosmopolites	1	1,3
Paleotropics	1	1,3
Pantropicals	2	2,6
Wide-ranging species	**4**	**5,1**
Afro-Malagasy	3	3,8
Afro-Tropicales	3	3,8
Multi-African	11	14,1
More common African species	**17**	**21,8**
Mountain women	37	47,4
Sudano-Zambéziennes	13	16,7
Species with regional distribution	**50**	**64,1**
L-SZ-G	4	5,1
Linking species	**4**	**5,1**
Endemic species	**1**	**1,3**
Guinean species	**2**	**2,6**
Total	78	100

Table 16: Selected groups of the montane base element at the Kaziramihunda 2 site (Mo: montane species; Mo(EA): East African montane species)

Groups	Number of groups	Raw spectrum (%)
Mo	23	62,2
Mo(EA)	14	37,8
Total	37	100

- *Ndubura 1 site*

At this site, montane species dominate, accounting for 39.1% of the total, followed by Sudano-Zambézian species (21.7%) and Pluri-African species (10.9%) (Table 17). These proportions show that Ndubura 1 is located in the montane zone, while the other phytogeographical elements are represented at low levels. The breakdown of species in the montane base element into typical montane groups (72.2%) and East African groups (27.8%) is shown in Table 18.

Table 17: Geographical distribution of species collected at the Ndubura 1

Phytogeographical types	Number of species	%
Cosmopolites	2	4,3
Pantropicals	2	4,3
Wide-ranging species	**4**	**8,7**
Afro-Malagasy	2	4,3
Afro-Tropicales	3	6,5
Multi-African	5	10,9
More common African species	**10**	**21,7**
Mountain women	18	39,1
Sudano-Zambéziennes	10	21,7
Species with regional distribution	**28**	**60,9**
L-SZ-G	3	6,5
Linking species	**3**	**6,5**
Endemic	**1**	**2,2**
Total	**46**	**100,0**

Table 18: Selected groups of the montane base element at the Ndubura 1 site (Mo: montane species; Mo(EA): East African montane species)

Groups	Number of groups	Raw spectrum (%)
Mo	13	72,2
Mo(EA)	5	27,8
Total	**18**	**100**

- *Ndubura 2 site*

Within the Ndubura 2 site, montane species dominate with 48.6% of the total, followed by Sudano-Zambézian species (10.8), pluri-African species (8.1%) and Afro-tropical species (8.1%), while the other phytogeographical elements are represented at a low rate (Table 19). The distribution of species in the montane base element into typical montane groups (61.1%) and East African groups (38.9%) is shown in Table 20. This geographical distribution places Ndubura 2 in the Afromontane domain.

Table 19: Geographical distribution of species collected at the Ndubura 2

Phytogeographical elements	Number of species	%
Paleotropics	1	2,7
Pantropicals	2	5,4
Wide-ranging species	**3**	**8,1**
Afro-Malagasy	1	2,7
Afro-Tropicales	3	8,1
Multi-African	3	8,1
More common African species	**7**	**18,9**
Mountain women	18	48,6
Sudano-Zambéziennes	4	10,8
Species with regional distribution	**22**	**59,5**
L-SZ-G	2	5,4
Linking species	**2**	**5,4**
Endemic species	**2**	**5,4**
Guinean species	**1**	**2,7**
Total	37	100,0

Table 20: Selected groups of the montane base element at the Ndubura 2 site (Mo: montane species; Mo(EA): East African montane species)

Groups	Number of groups	Raw spectrum (%)
Mo	11	61,1
Mo(EA)	7	38,9
Total	18	100

- *Ndubura 3 site*

Analysis of the geographical distribution of species at Ndubura 3 shows that montane species top the list with 60.5%, placing the site in the afromontane domain. They are followed by Sudano-Zambézian species (14.0%) and Pluri-African species (9.3%). The others are poorly represented (Table 21). Montane species are subdivided into the omni-montane group (61.5%) and the East African montane group (38.5%) (Table 22).

Table 21: Geographical distribution of species collected at the Ndubura 3

Phytogeographical elements	Number of species	%
Paleotropics	1	2,3
Pantropicals	1	2,3
Wide-ranging species	**2**	**4,7**
Afro-Malagasy	1	2,3
Afro-Tropicales	3	7,0
Multi-African	4	9,3
More common African species	**8**	**18,6**
Mountain women	26	60,5
Sudano-Zambéziennes	6	14,0
Species with regional distribution	**32**	**74,4**
L-SZ-G	1	2,3
Linking species	**1**	**2,3**
Total	**43**	**100**

Table 22: Selected groups of the montane base element at the Ndubura 3 site (Mo: montane species; Mo(EA): East African montane species)

Groups	Number of groups	Raw spectrum (%)
Mo	16	61,5
Mo(EA)	10	38,5
Total	26	100

IV.1.4. Structural features

- **Physionomy of different sites**

Generally speaking, five strata can be distinguished in the Teza sector of KNP: the upper tree stratum (30-50 m), the middle tree stratum (20-30 m), the lower tree stratum (7-20 m), the shrub stratum (2-7 m) and the sub-shrub and/or herbaceous stratum (0-2 m). The cover and characteristic species of each stratum vary according to the study sites.

- *Kaziramihunda 1 site*

Four strata stand out at the Kaziramihunda 1 site: the medium tree stratum (20-30 m) is represented by *Polyscias fulva, Macaranga kilimandscharica, Tabernaemontana johnstonii, Sapium ellipticum, Syzygium parvifolium* and *Magnistipula butayei*. The average cover of this stratum is estimated at 70%, with an average height of 25 m.

The lower tree stratum (7-20 m) is marked by *Galiniera saxifraga,*

Tabernaemontana johnstonii, Myrianthus holstii, Xymalos monospora, Syzygium parvifolium, Macaranga kilimandscharica, Bersama abyssinica, Strombosia scheffleri, Symphonia globulifera, etc., with an average height of 13 m. Average cover is estimated at 80%. The two preceding strata prevent the penetration of a large amount of light down to ground level, resulting in the specific poverty of the lower strata.

The shrub layer (2-7 m) mainly comprises *Lindackeria kivuensis, Cyathea manniana, Alchornea hirtella, Rauvolfia mannii, Dracaena afromontana, Chassalia subochreata, Tabernaemontana johnstonii, Myrianthus holstii, Galiniera coffeoides, Macaranga kilimandscharica, Strombosia scheffleri, Symphonia globulifera,* etc., with an average height of 5 m and average cover of 30%.

With an average cover of 10%, the sub-shrub and/or herbaceous stratum (0-2 m) is smothered by the three preceding strata due to light deprivation and abundant litter. Average height is estimated at 0.7 m. Species such as *Sericostachys scandens, Xymalos monospora, Syzygium parvifolium, Strombosia scheffleri, Galiniera coffeoides, Asplenium elliotii, Asplenium sandersonii, Kalanchoe crenata, Dracaena afromontana, Impatiens burtonii, Impatiens purpureo-violacea, Impatiens stuhlmannii, Symphonia globulifera,* etc, represent this stratum.

Located at an altitude of 2128 m, the Kaziramihunda 1 site is part of the middle horison. The upper tree stratum, consisting of giant trees, has been eliminated, and trees such as *Entandrophragma excelsum, Prunus africana* and *Parinari excelsa,* which have been reported in this horizon, are virtually non-existent in this plant formation. It is therefore a degraded mountain rainforest. Figure 16 shows the physiognomy of the Kaziramihunda 1 site.

Fig. 16: Physiognomic aspect of the Kaziramihunda 1 site. Photo: D. Nzoyisaba, 2013

- *Rwahirirwa site*

Structural analysis of this site shows the existence of four strata: the upper tree stratum (30-50 m) is essentially represented by a single species, *Syzygium parvifolium*. Coverage of this stratum is estimated at 20%, with a height of 35 m.

The lower tree stratum (7-20 m) reaches a height of 14 m and covers 50%. It is mainly composed of *Symphonia globulifera, Galiniera coffeoides, Psychotria palustris, Xymalos monospora, Syzygium parvifolium, Macaranga kilimandscharica, Dracaena afromontana, Coccinia mildbraedii* and others. This stratum is also home to lianas, mainly *Schefflera goetzenii, Schefflera abyssinica, Embelia schimperi* and *Embelia libeniana*, and many epiphytes.

The shrub layer (2-7 m) mainly comprises *Psychotria palustris, Alchornea hirtella, Virectaria major, Allophylus chaunostachys, Chassalia subochreata,*

Embelia Schimperi, Schefflera abyssinica, Schefflera goetzenii, Dracaena afromontana, Symphonia globulifera, etc., with an average height of 5 m and average cover of 15%. At this site, *Mimulopsis solmsii* forms clumps in some places reaching a height of 5 m.

The shrub and/or herb layer (0-2 m) is mainly composed *of Asplenium aethiopicum, Asplenium elliotii, Rourea thomsonii, Impatiens burtonii, Impatiens purpureo- violacea, Virectaria major, Syzygium parvifolium, Cyperus* sp, *Begonia meyeri-johannis, Smilax anceps, Tabernaemontana johnstonii* etc., interspersed with young Rubiaceae, notably *Chassalia subochreata* and *Galiniera coffeoides*, less than 2 m high. The average cover of this stratum is estimated at 80%.

The Rwahirirwa site is part of the middle horizon at 2192 m altitude. It is a secondary mountain forest characterized by the frequent presence of *Syzygium parvifolium* and *Macaranga kilimandscharica*. Figure 17 shows the physiognomy of the Rwahirirwa site.

Fig. 17: Physiognomic aspect of the Rwahirirwa site. Photo: D. Nzoyisaba, 2013

- Kaziramihunda 2 site

This site comprises five strata, with many large trees whose crowns let in a great deal of light, thus promoting floristic diversity in the herbaceous stratum. This good light penetration also justifies the high floristic richness compared with other sites.

The upper tree stratum (30-50 m) includes *Prunus africana, polyscias fulva* and *syzygium parvifolium*, with an average height of 35 m and an estimated crown cover of 40%.

The medium tree stratum (20-30 m) is mainly composed of *Prunus africana, Polyscia fulva, Syzygium parvifolium, Tabernaemontana johnstonii, Myrianthus holstii* and a few lianas such as *Schefflera goetzenii, Urera hypselodendron, Mimosa montana* and many others, with a 50% cover. The average height of this stratum is 25 m.

The lower tree stratum (7-20 m) reaches a height of 13 m and has a crown cover of 25%. It is mainly composed of *Myrianthus holstii, Tabernaemontana johnstonii, Xymalos monospora, Prunus africana, Dracaena afromontana, Syzygium parvifolium, Magnistipula butayei*, etc.

The shrub stratum, up to 4 m high with a 25% cover, is represented by *Piper capense, Chassalia subochreata, Symphonia globulifera, Galiniera coffeoïdes, Strombosia scheffleri, Dracaena afromontana, Myrianthus holstii, Psychotria palustris, Clerodendrum Sp*, etc.
The shrub and/or herbaceous stratum, averaging 0.8 m in height, is diverse but less dense, with a cover of around 50%, and mainly comprises *Chassalia subochreata, Rourea thomsonii, Gynura scandens, Myrianthus holstii, Sericostachys scandens, Urera hypselodendron, Symphonia globulifera, Galiniera coffeoïdes, Setaria megaphylla, Asplenium dregeanum, Asplenium elliotii, Asplenium glamosum.*

Located at an altitude of 2169 m, Kaziramihunda 2 belongs to the middle horizon. Giant *Prunus africana* trees appear in the upper tree stratum. This is a primary mountain forest, but the absence of *Entandrophragma excelsum* and *Parinari excelsa* in the upper tree stratum may indicate some degradation. Figure 18 shows the physiognomy of the Kaziramihunda 2 site.

Fig. 18: Physiognomic aspect of the Kaziramihunda 2 site. Photo: D. Nzoyisaba, 2013

- *Ndubura 1 site*

The Ndubura 1 site is sadratified. The upper and middle tree strata are not represented. The lower tree stratum (7-20 m) is made up of trees reaching an average height of 15 m and is less diverse, with only
2 species *Neoboutonia macrocalyx* and *Dombeya goetzenii*. Average cover is estimated at 1%.

The shrub layer (2-7 m) is less dense (5% cover), with a few *Brillantaisia cicatricosa, Gynura scandens, Rubus pinnatus, Clerodendrum johnstonii, Discopodium penninervium, Triumfetta cordifolia* and *Neoboutonia macrocalyx. Ensete ventricosum* is one of the site's characteristic species, with giant individuals reaching 3 m in height. This stratum reaches an average height of 3 m.

The shrub and/or herb layer (0-2 m) is dense but also diversified (average cover estimated at 100%). It reaches an average height of 1.5 m. It is mainly represented by *Anisosepalum humbertii, Solanum nigrum, Urera hypselodendron, Piper capense, Brillantaisia cicatricosa, Gynura scandens, Rubus pinnatus, Pteridium aquilinum,* etc.

Ndubura 1 is described as a pre-forest scrubland. Figure 19 shows the physiognomy of the Ndubura 1 site.

Fig.19: Physiognomic aspect of the Ndubura 1 site. Photo: D. Nzoyisaba, 2013

- *Ndubura 2 site*

Five strata have been distinguished at this site. The upper tree stratum (30-50 m) includes *Chrysophyllum gorungosanum, Strombosia scheffleri* and *Synsepalum attenuatum*, with an average height of 35 m and an estimated crown cover of 50%.

The medium tree stratum (20-30 m) is mainly composed of *Macaranga kilimandscharica, Chrysophyllum gorungosanum, Myrianthus holstii, Tabernaemontana johnstonii, Strombosia scheffleri* and *Polyscias fulva*, with a cover of 30%. Species in this stratum reach an average height of 25 m.

The lower tree stratum (7-20 m) reaches a height of 12 m and covers 30% of the plant. It is mainly composed of *Myrianthus holstii, Xymalos monospora, Symphonia globulifera, Alchornea hirtella, Tabernaemontana johnstonii, Macaranga kilimandscharica, Chrysophyllum gorungosanum, Strombosia scheffleri* and *Synsepalum attenuatum*.

The shrub stratum, up to 4 m high with a coverage of 15%, is represented by *Strombosia scheffleri, Tabernaemontana johnstonii, Symphonia globulifera, Chassalia subochreata, Alchornea hirtella, Macaranga kilimandscharica, Dracaena afromontana, Rauvolfia mannii*, etc.

The shrub and/or herb layer, averaging 0.5 m in height, is less dense and diversified, with a cover of around 15%. The tree crowns do not allow much light to penetrate to the ground, which makes this stratum increasingly thin. It mainly comprises *Alchornea hirtella, Sericostachys scandens, Gynura scandens, Myrianthus holstii, Chrysophyllum gorungosanum, Chassalia subochreata, Asplenium elliotii, Asplenium mannii, Dracaena afromontana* and *Galiniera coffeoides*.

Ndubura 2 is part of the upper horizon and can be described as primary mountain forest. Figure 20 shows the physiognomy of the Ndubura 2 site.

Fig. 20: Physiognomic aspect of the Ndubura 2 site. Photo: D. Nzoyisaba, 2013

- *Ndubura 3 site*

This site has a low herbaceous layer and is made up of trees of a very high density.
considerable height. Five strata have been identified:

The upper tree stratum (30-50 m) is marked by large trees of *Symphonia globulifera*, *Strombosia scheffleri* and *Chrysophyllum gorungosanum* with circumferences of up to 308 cm. Tree crown cover is 65%, with an average height of 35 m.

The medium tree stratum (20-30 m) is dominated by *Macaranga kilimandscharica*, *Chrysophyllum gorungosanum*, *Strombosia scheffleri*, *Tabernaemontana johnstonii*, *Xymalos monospora*, *Myrianthus holstii* and *Symphonia globulifera*, with an average height of 25 m and an estimated crown cover of 50%.

The lower tree stratum (7-20 m) consists mainly of *Macaranga kilimandscharica*, *Neoboutonia macrocalyx*, *Tabernaemontana johnstonii*, *Chrysophyllum gorungosanum*, *Myrianthus holstii*, *Symphonia globulifera* and

Xymalos monospora, with an estimated average cover of 20%. The average height of this stratum is estimated at 15 m.

The shrub layer (2-7 m), up to 5 m high, is dominated by *Chrysophyllum gorungosanum, Myrianthus holstii, Symphonia globulifera, Strombosia scheffleri, Chassalia subochreata* and *Xymalos monospora.* It is less diverse, with an average cover of 60%.

The shrub and/or herb layer (0-2 m) averages 0.8 m in height. The dominant species in this stratum are: *Mimulopsis solmsii, Asplenium dregeanum, Asplenium elliotii, Asplenium glamosum, Hypoestes forskalei, Sericostachys scandens, Chassalia subochreata, Rauvolfia manni, Strombosia scheffleri* and *Tabernaemontana johnstonii.* The average cover of this stratum is estimated at 30%.

This physiognomy allows us to classify Ndubura 3 as a primary mountain forest. Figure 21 shows the physiognomy of the Ndubura 3 site.

Fig. 21: Physiognomic aspect of the Ndubura 3 site. Photo: D. Nzoyisaba, 2013

- **Distribution of stems by circumference class**

- *Kaziramihunda 1 site*

At this site, circumference measurements were taken on 278 stems, of which 103 stems (37%) had circumferences concentrated in the 15-24 cm range. In this interval, *Tabernaemontana johnstoni* predominates with 22 stems. The 25-34 cm interval features 68 stems, of which 27 stems (39.7%) are *Macaranga kilimandscharica*. The class with the highest circumferences is [215-225[, represented by one individual of *Polyscias fulva*. The coefficient of determination (R^2 =0.89) is very high, which could be due to the existence of natural conditions favorable to forest regeneration in recent years for the species observed. This shows that the site is stable.

The number of stems per hectar is highest in Kaziramihunda 1 (2317 stems/ha), indicating that this is a dense forest. Figure 22 shows the distribution of the number of stems per circumference class. Details of the species and number of stems measured per class are shown in Table 1 in Appendix 3.

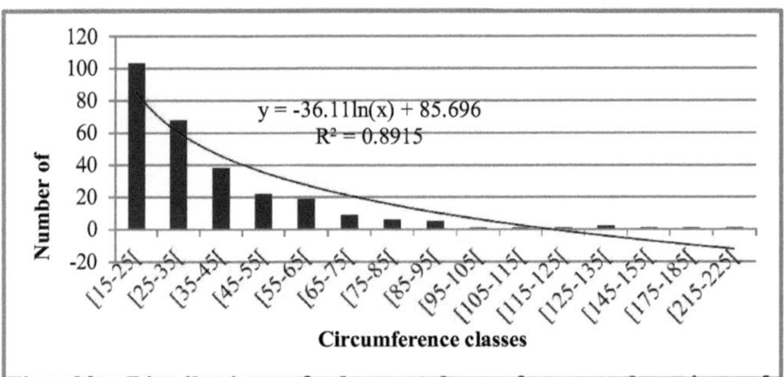

Fig. 22: Distribution of the number of stems by circumference class at the Kaziramihunda 1 site

- Rwahirirwa site

Out of 147 stems measured, 41 stems, or 28%, have a circumference between 35 and 44 cm. The [15-25[class has only 27 stems, while the class with the highest circumference is [265-275[, represented by a single *Syzygium parvifolium*. The coefficient of determination R^2 is equal to 0.71, showing that the site is stable. Deadwood cuttings of around 2 m^2 were observed. The number of stems is estimated at 1,225 stems/ha, showing that Rwahirirwa is less dense than Kaziramihunda 1. Figure 23 shows the distribution of the number of stems by circumference class; details of the species and number of stems per circumference class are shown in Table 2 in Appendix 3.

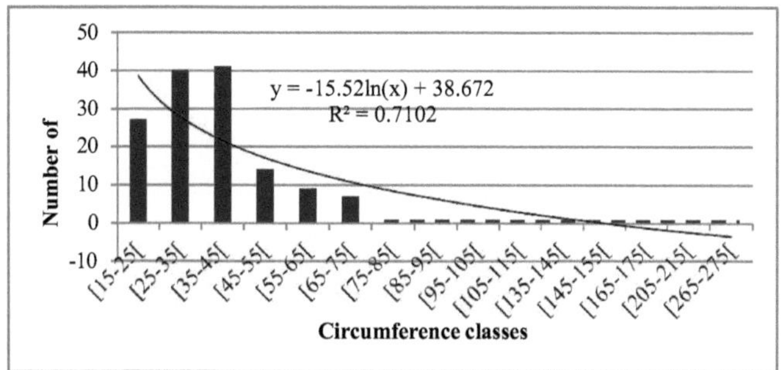

Fig. 23: Distribution of the number of stems by circumference class at the Rwahirirwa site

- *Kaziramihunda 2 site*

The circumference class distribution of stems measured at the Kaziramihunda 2 site is shown in figure 24. Of 170 stems measured, 60 stems, or 35.3%, have a circumference ranging from 15 to 24 cm. The class with the largest circumference is [415-425[, represented by a single *Prunus africana*. The stability of the site is reflected in the high value of the coefficient of determination (R^2 = 0.8). Good light penetration favours germination of seeds from mature trees. With 1063 stems per hectar, Kaziramihunda 2 is also dense. However, Kaziramihunda 2 is less dense than Kaziramihunda 1 and Rwahirirwa.

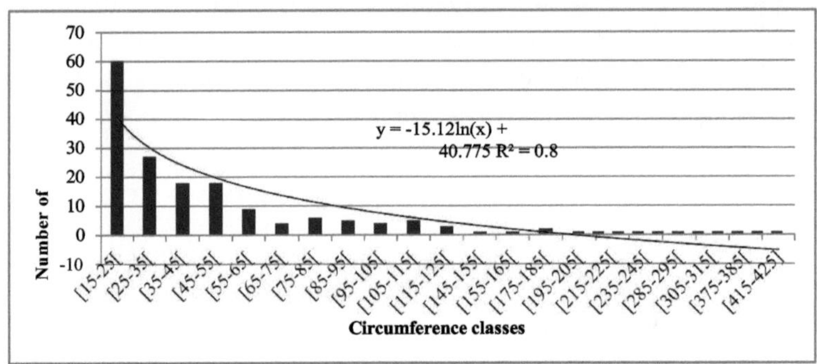

Fig. 24: Distribution of the number of stems by circumference class at the Kaziramihunda 2 site

- Ndubura 1 site

With a few individuals of *Neoboutonia macrocalyx*, *Discopodium penninervium* and *Dombeya goetznii*, this was the first site to record fewer stems measured (31 stems). The most represented class is [15-25[with 15 stems, or 48.4%.

The class with the highest circumference is [85-95[, represented by two individuals of *Neoboutonia macrocalyx*. The coefficient of determination R^2 equal to 0.6 testifies to the instability of the site's vegetation. Indeed, with the elimination of the three species mentioned above, the site will be purely grassy. The number of stems is estimated at 344 stems/ha, making Ndubura the least dense of the sites. The distribution in circumference classes of stems measured at the Ndubura 1 site is shown in figure 25.

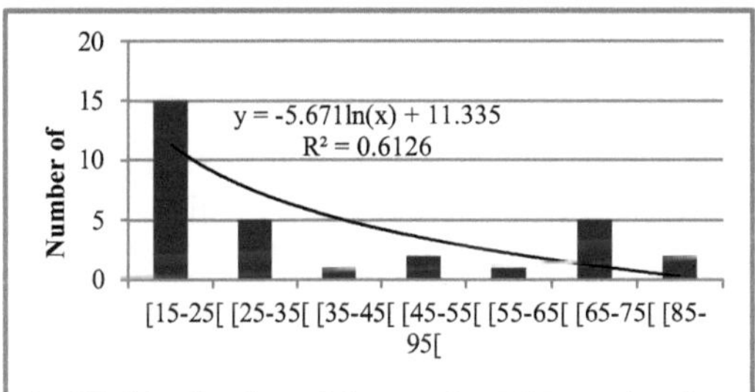

Fig. 25: Distribution of the number of stems by circumference class at the Ndubura1

- *Ndubura 2 site*

Figure 26 shows the distribution of the number of stems by circumference class. Out of a total of 108 stems measured, 66 stems, or 61%, ranged in circumference from 15 to 34 cm. The [15-25[class alone accounts for 50% of the stems measured. The classes with trees with the highest circumferences are [255-265[and [265- 275[, each represented by 1 *Chrysophyllum gorungosanum* individual. The coefficient of determination R^2 is equal to 0.55, showing that the vegetation on the site is not stable. The number of stems was estimated at 900 stems/ha, showing that Ndubura 2 is less dense.

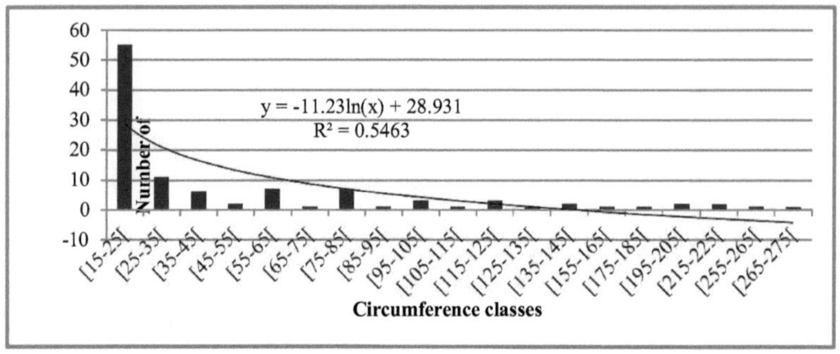

Fig. 26: Distribution of the number of stems by circumference class at the Ndubura 2

- *Ndubura 3 site*

The circumference class distribution of stems measured at the Ndubura 3 site is shown in figure 27. Of 97 stems measured, 48 stems (49.5%) had a circumference ranging from 15 to 34 cm. The class with the largest circumference is [305-315[, represented by a single *Symphonia globulifera*. The coefficient of determination R^2 is 0.75, showing the stability of the site. The number of stems was estimated at 1078 stems/ha, showing that Ndubura 3 is denser than Ndubura 2, Ndubura 1 and Kaziramihunda 2, but less dense than Kaziramihunda 1 and Rwahirirwa.

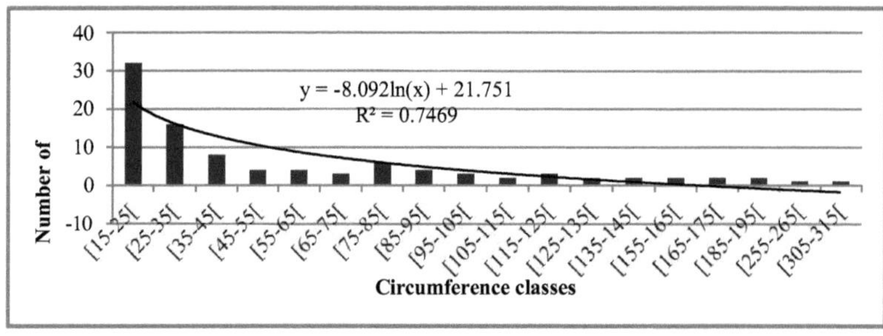

Fig. 27: Distribution of the number of stems by circumference class at the Ndubura 3

- **Basal area**

Basal area varies from site to site, depending on the number and circumference of trees measured. Sites with large stems, such as Kaziramihunda 2, Ndubura 2 and Ndubura 3, have high basal areas. Ndubura 1 recorded a low basal area value of 5.2 m²/ha.

- *Kaziramihunda 1 site*

Table 23 shows the basal area in m²/ha of the stems measured. The basal area of the Kaziramihunda 1 site is 38 m²/ha, indicating that it is a dense humid forest. *Macaranga kilimandscharica* leads with 23.8% of the total basal area, followed by *Xymalos monospora* with 14.9%, *Polyscias fulva* with 14.6% and *Tabernaemontana johnstonii* with 13.6%. Other species are poorly represented.

Table 23: Basal area in m² /ha by circumference class at the Kaziramihunda 1 site

Circumference classes	Macaranga kilimandscharica	Tabernaemontana johnstonii	Xymalos monospora	Galiniera coffeoides	Psychotria palustris	Magnistipula butayei	Rytigynia Kivuensis	Dracaena afromontana	Bersama abyssinica
[15-25[	391,8	709,8	315,5	574,8	231,8	77,4	223,9	31,8	45,9
[25-35[	1757,1	1299,3	517,4	214,9	99,5	81,5	67,0	76,5	0,0
[35-45[	1270,6	859,9	1242,2	287,6	254,8	274,2	0,0	127,4	133,8
[45-55[	1781,3	1344,2	0,0	0	0	168,5	215,3	207,0	,0
[55-65[	1856,1	1194,7	927,95	0	249,7	0	0	554	0
[65-75[	1582,9	802,7	368,15	0	0	0	0	0	390
[75-85[	561,78	0	496,89	548,5	0	472	0	0	0
[85-95[	630,65	0	1906	0	0	0	0	0	0
[95-105[	0	0	0	0	0	0	0	0	0
[105-115[	1016,6	0	0	0	0	0	0	0	0
[115-125[	0	0	1052,9	0	0	0	0	0	0
[125-135[	0	0	0	0	0	1387,3	0	0	0
[145-155[	0	0	0	0	0	0	0	0	0
[175-185[	0	0	0	0	0	0	0	0	0
[215-225[	0	0	0	0	0	0	0	0	0
T0tal (Cm)²	10848,9	6210,5	6827,0	1625,8	835,8	2460,9	506,2	997,0	569,7
Gi (m² /ha)	9,04	5,18	5,69	1,35	0,70	2,05	0,42	0,83	0,47
%	23,791	13,62	14,971	3,5654	1,833	5,3967	1,11	2,19	1,249

Circumference classes	Polyscias fulva	Lindackeria kivuensis	Chassalia subochreata	Indét	Shirakiopsis elliptica	Myrianthus holstii	Syzygium parvifolium	Cyathea sp	Total in Cm²	Gi in m /ha²	%
[15-25[	0	46	20,4	0	0	133,8	92,3	87,9	2983,1	2,49	6,5
[25-35[	0	0	0	71,7	0	178,7	76,5	219,7	4659,8	3,88	10,2
[35-45[	0	0	0	0	103,2	109,0	0	0	4662,8	3,89	10,2
[45-55[	0	0	0	0	0	161,2	161,2	0	4038,6	3,37	8,9
[55-65[	0	0	0	0	0	592,5	0	0	5375,3	4,48	11,8
[65-75[	0	0	0	0	0	346,8	0	0	3490,5	2,91	7,7
[75-85[	0	0	0	0	0	472	484,4	0	3035,6	2,53	6,7
[85-95[	0	0	0	0	0	0	588,9	0	3125,6	2,60	6,9
[95-105[	0	0	0	0	0	0	844,7	0	844,7	0,70	1,9
[105-115[	0	0	0	0	0	0	0	0	1016,6	0,85	2,2
[115-125[	0	0	0	0	0	0	0	0	1052,9	0,88	2,3
[125-135[	0	0	0	0	0	0	1429,6	0	2816,9	2,35	6,2
[145-155[	0	0	0	0	1864	0	0	0	1863,8	1,55	4,1
[175-185[	2695,5	0	0	0	0	0	0	0	2695,5	2,25	5,9
[215-225[	3959,3	0	0	0	0	0	0	0	3959,3	3,30	8,7
T0tal (Cm)²	6654,8	46,0	20,4	71,7	1967,0	1994,0	3677,6	307,6	45620,9		
Gi (m² /ha)	5,55	0,04	0,02	0,06	1,64	1,66	3,06	0,26		38,0	
%	14,594	0,1	0,04	0,16	4,314	4,373	8,0649	0,67			100

- *Rwahirirwa site*

Table 24 shows the basal area in m²/ha of stems measured at the Rwahirirwa site. The basal area at the Rwahirirwa site is

74

26.8 m²/ha. *Syzygium parvifolium* tops the list with 39.6% of total basal area, followed by *Macaranga kilimandscharica* with 20.2% and *Galiniera coffeoides* with 10%. Other species are poorly represented.

Table 24: Basal area in m² /ha by circumference class at the Rwahiritwa site

Circumference classes	*Galiniera coffeoides*	*Macaranga kilimandscharica*	*Xymalos monospora*	*Syzygium parvifolium*	*Psychotria palustris*	*Dracaena afromontana*	Indét	*Tabernaemontana johnstonii*	*Symphonia globulifera*	*Maesa lanceolata*	*Myrianthus holstii*	*Bersama abyssinica*	*Cyathea maniana*	*Total in Cm²*	*Gi in m /ha²*	%
[15-25[	394,8	105,3	83,2	35,1	151	0	0	0	0	0	0	0	25,8	795,6	0,7	2,5
[25-35[	816,8	532,1	134	0	81,5	0	0	0	58	0	49,8	0	1300,7	2972,9	2,5	9,3
[35-45[	1804	611,7	472	97,5	0	0	0	0	0	0	0	0	1713,9	4698,6	3,9	14,6
[45-55[	215,3	879,4	964	207	0	0	0	0	223,6	207,1	0	0	0	2696	2,2	8,4
[55-65[	0	1986	250	0	0	0	0	0	0	0	0	296,3	0	2532,4	2,1	7,9
[65-75[	0	1126	726	0	0	368,2	0	368,2	0	0	0	0	0	2587,9	2,2	8,1
[75-85[	0	535,4	0	0	0	0	0	0	0	0	0	0	0	535,4	0,4	1,7
[85-95[	0	0	0	703,5	0	0	0	0	0	0	0	0	0	703,5	0,6	2,2
[95-105[	0	718,6	0	0	0	0	0	0	0	0	0	0	0	718,6	0,6	2,2
[105-115[	0	0	0	877,8	0	0	0	0	0	0	0	0	0	877,8	0,7	2,7
[135-145[	0	0	0	1605	0	0	0	0	0	0	0	0	0	1605,4	1,3	5,0
[145-155[	0	0	0	0	0	0	0	0	0	0	0	0	0	0	0,0	0,0
[165-175[	0	0	0	0	0	0	2194	0	0	0	0	0	0	2194	1,8	6,8
[205-215[	0	0	0	3478	0	0	0	0	0	0	0	0	0	3477,8	2,9	10,8
[265-275[	0	0	0	5719	0	0	0	0	0	0	0	0	0	5718,5	4,8	17,8
T0tal (Cm)²	3230	6495	2628	12723	233	368,2	2194	368,2	281,6	207,1	49,8	296,3	3040,4	32114,4		
Gi (m² /ha)	2,692	5,412	2,19	10,6	0,19	0,307	1,83	0,307	0,235	0,173	0,04	0,247	2,5337		26,8	
%	10,04	20,2	8,17	39,56	0,72	1,145	6,82	1,145	0,876	0,644	0,15	0,921	9,454			100

- ***Kaziramihunda 2 site***

The basal area of Kaziramihunda 2 is shown in Table 25. *Prinus africana* dominates with 30.84% of the total basal area at this site, followed by *Tabernaemontana johnstonii* with 17.55% and *Syzygium parvifolium* with 14.12%. The others account for a small proportion of basal area. Kaziramihunda 2 has a basal area of 57.22 m²/ha.

Table 25: Basal area in m² /ha by circumference class at the Kaziramihunda 2 site

Circumference classes	*Myrianthus holstii*	*Tabernaemontana johnstonii*	*Xymalos monospora*	*Dracaena afromontana*	*Magnistipula butayei*	*Acacia montigena*	*Chassaria subochreata*	*Rytigynia kivuensis*	*Prinus africana*	Indet 1	*Polyscias fulva*	*Schefflera abyssinica*	Indet	*Syzygium parvifolium*	*Pauridiantha paucinervis*	*Alchornea hirtella*	*Cyathea maniana*	Total (cm)²	Gi (m²/ha)	%
[15-25[	597,2	159,2	202,7	229	54,5	38,5	143,5	217,9	0	20,38	0	0	0	0	0	38,5	0	1701,38	1,06	1,86
[25-35[	787	62,4	227,9	174,1	0	76,5	0	165,2	0	0	0	81,5	66,95	0	0	0	92	1733,55	1,08	1,89
[35-45[	853,7	515,9	543,2	236	0	0	0	0	0	0	0	0	0	0	140,4	0	0	2289,20	1,43	2,50
[45-55[	1417	344,3	1306	175,9	0	0	0	0	0	0	0	168,5	0	0	0	0	0	3411,10	2,13	3,73
[55-65[	937,8	1089,8	563,7	0	0	0	0	0	0	0	0	0	0	0	0	0	0	2591,30	1,62	2,83
[65-75[	0	1238,2	0	401,3	0	0	0	0	0	0	0	0	0	0	0	0	0	1639,50	1,02	1,79
[75-85[	993,8	548,5	1019	0	0	0	0	0	0	0	0	0	0	0	561,8	0	0	3123,20	1,95	3,41
[85-95[	1875	703,5	0	0	0	0	0	0	0	0	0	0	0	575,2	0	0	0	3153,60	1,97	3,44
[95-105[	1641	844,6	0	844,7	0	0	0	0	0	0	0	0	0	0	0	0	0	3329,80	2,08	3,64
[105-115[	877,8	1891,9	998,7	1017	0	0	0	0	0	0	0	0	0	0	0	0	0	4785,00	2,99	5,23
[115-125[	2255	1224,2	0	0	0	0	0	0	0	0	0	0	0	0	0	0	0	3479,10	2,17	3,80
[145-155[	0	1863,8	0	0	0	0	0	0	0	0	0	0	0	0	0	0	0	1863,80	1,16	2,04
[155-165[	0	0	0	0	0	0	0	0	0	0	0	0	0	1987,6	0	0	0	1987,60	1,24	2,17

[175-185[	0	2551	0	0	0	0	0	0	2637,2	0	0	0	0	0	0	0	0	5188,20	3,24	5,67
[195-205[	0	3027,5	0	0	0	0	0	0	0	0	0	0	0	0	0	0	0	3027,50	1,89	3,31
[215-225[	0	0	0	0	0	0	0	0	0	0	0	0	0	3853,5	0	0	0	3853,50	2,41	4,21
[235-245[	0	0	0	0	0	0	0	0	0	0	4624,3	0	0	0	0	0	0	4624,30	2,89	5,05
[285-295[	0	0	0	0	0	0	0	0	0	0	0	0	0	6512,4	0	0	0	6512,40	4,07	7,11
[305-315[	0	0	0	0	0	0	0	0	0	0	7651,3	0	0	0	0	0	0	7651,30	4,78	8,36
[375-385[	0	0	0	0	0	0	0	0	11557,4	0	0	0	0	0	0	0	0	11557,40	7,22	12,62
[415-425[	0	0	0	0	0	0	0	0	14044,5	0	0	0	0	0	0	0	0	14044,50	8,78	15,34

Total (cm)²	12234	16065	4861	3078	54.5	115	143.5	383.1	28239.1	20.38	12275.6	250	66.95	12929	702.2	38.5	92	91547.23
Gi (m/ha²)	7.65	10.04	3.04	1.92	0.03	0.07	0.09	0.24	17.65	0.01	7.67	0.16	0.04	8.08	0.44	0.02	0.06	57.22
%	13.36	17.55	5.31	3.36	0.06	0.13	0.16	0.42	30.84	0.02	13.41	0.27	0.07	14.12	0.77	0.04	0.10	100

- Ndubura 1 site

Table 26 shows the basal area in m²/ha of stems measured at the Ndubura 1 site. The basal area of the site is 5.2 m²/ha, the lowest of the other sites. The site could be considered as a wasteland belonging to the pre-forest recrûs. Only 3 species were measured at this site: *Neoboutonia macrocalyx* with 80.43% of total basal area, *Dombeya goetznii* with 14.19% and *Discopodium penninervium* with 5.4%.

Table 26: Basal area in m² /ha by circumference class at the Ndubura 1 site

Circumference classes	*Neoboutonia macrocalyx*	*Dombeya goetznii*	*Discopodium penninervium*	Total (cm)²	Gi (m /ha²	%
[15-25[	114,9	0	253,1	368	0,41	7,86
[25-35[	258,7	71,7	0	330,4	0,37	7,06
[35-45[	103,2	0	0	103,2	0,11	2,21
[45-55[	168,5	191,2	0	359,7	0,40	7,69
[55-65[	306,1	0	0	306,1	0,34	6,54
[65-75[	1551,6	401,35	0	1952,95	2,17	41,73
[85-95[	1261,3	0	0	1261,3	1,40	26,95
Total (cm)²	3764,3	664,25	253,1	4681,7		
Gi (m /ha²	4,18	0,74	0,28		5,20	
%	80,43	14,19	5,41			100

- Ndubura 2 site

Table 27 shows the basal area in m²/ha of stems measured at the Ndubura 2 site, with a total basal area of 39.5 m²/ha. *Chrysophyllum gorungosanum* tops the list with 40.07% of the total basal area, followed by *Macaranga kilimandscharica* with 10.94% and *Polyscias fulva* with 8%. Other species are poorly represented.

Table 27: Basal area in m² /ha by circumference class at the Ndubura 2 site

Circumference classes	*Macaranga kilimandscharica*	*Myrianthus holstii*	*Tabernaemontana johnstonii*	*Pauridiantha paucinervis*	*Chassaria subochreata*	*Alchornea hirtella*	*Chrysophyllum gorungosanum*	*Strombosia Scheffleri*	*Xymalos monospora*	*Snsepalum seretii*	*Magnistipula butayei*	*Rauvolfia mannii*	*Dracaena afromontana*	*Psychotria palustris*	*Sericanthe burundensis*	*Symphonia globulifera*	ndét	*Polyscia fulva*	Total (cm)²	Gi (m /ha²)	%
[15-25[	261,7	28,7	66,9	0	46	714,3	0	0	20,4	0	42,1	20,4	79,9	57,5	74,8	73,6	0	0	1486,3	1,2	3,14
[25-35[	294,5	0	81,5	0	0	187,3	0	0	148,1	0	0	0	0	0	0	0	0	0	711,4	0,6	1,50
[35-45[	401,5	0	0	103,2	0	0	0	0	0	0	0	0	108,9	0	0	108,9	0	0	722,5	0,6	1,52
[45-55[	0	199	0	168,5	0	0	0	0	0	0	0	0	0	0	0	0	0	0	367,5	0,3	0,78
[55-65[	326,1	757,1	267,8	0	0	0	277,1	296,3	0	0	0	0	0	0	0	0	0	0	1924,4	1,6	4,06
[65-75[	0	0	0	0	0	0	0	401,4	0	0	0	0	0	0	0	0	0	0	401,4	0,3	0,85
[75-85[	1044,7	0	1096,9	0	0	0	0	0	447,9	484,4	0	0	496,9	0	0	0	0	0	3570,8	3,0	7,53
[85-95[	0	0	0	0	0	0	0	616,6	0	0	0	0	0	0	0	0	0	0	616,6	0,5	1,30
[95-105[	844,7	812,2	0	0	0	0	0	812,2	0	0	0	0	0	0	0	0	0	0	2469,1	2,1	5,21
[105-115[	0	0	0	0	0	0	0	1034,7	0	0	0	0	0	0	0	0	0	0	1034,7	0,9	2,18
[115-125[	0	0	1204,5	0	0	0	0	1052,9	0	0	0	0	0	0	0	0	1165,7	0	3423,1	2,9	7,22
[125-135[	0	0	0	0	0	0	0	0	0	1345,5	0	0	0	0	0	0	0	0	1345,5	1,1	2,84
[135-145[	0	1582,8	0	0	0	0	1650,9	0	0	0	0	0	0	0	0	0	0	0	3233,7	2,7	6,82
[155-165[	2012,8	0	0	0	0	0	0	0	0	0	0	0	0	0	0	0	0	0	2012,8	1,7	4,25
[175-185[	0	0	0	0	0	0	2579,6	0	0	0	0	0	0	0	0	0	0	0	2579,6	2,1	5,44
[195-205[	0	0	0	0	0	0	3089,9	0	0	3248,7	0	0	0	0	0	0	0	0	6338,6	5,3	13,37
[215-225[	0	0	0	0	0	0	0	0	0	0	0	0	0	0	0	0	0	3818,6	3818,6	3,2	8,06
[255-265[	0	0	0	0	0	0	5507,1	0	0	0	0	0	0	0	0	0	0	0	5507,1	4,6	11,62
[265-275[	0	0	0	0	0	0	5890,4	0	0	0	0	0	0	0	0	0	0	0	5890,4	4,9	12,43

Total (cm)²	5186	3379,8	2717,6	271,7	46	901,6	18995	3401,9	1429	5078,6	42,1	20,4	685,7	57,5	74,8	182,5	1165,7	3818,6	47454	
Gi (m /ha²	4,32	2,82	2,26	0,23	0,04	0,75	15,83	2,83	1,19	4,23	0,04	0,02	0,57	0,05	0,06	0,15	0,97	3,18	39,5	
%	10,94	7,13	5,73	0,57	0,10	1,90	40,07	7,18	3,01	10,71	0,09	0,04	1,45	0,12	0,16	0,39	2,46	8,06		100

- *Ndubura 3 site*

The basal area of Ndubura 3 is shown in Table 28. *Chrysophyllum* gorungosanum dominates with 24.96% of the total basal area at this site, followed by *Myrianthus holstii* with 15.8% and *Symphonia globulifera* with 15%. The others account for a small proportion of basal area. Measured basal area is 58.03 m² /ha.

Table 28: Basal area in m² /ha by circumference class at the Ndubura 3 site

Classes of circumference	*Macaranga kilimandscharica*	*Chrysophyllum gorungosanum*	*Myrianthus holstii*	*Strombosia scheffleri*	*Erythrococca bongensis*	*Psychotria palustris*	*Rytigynia bridsonii*	*Garcinia volkensii*	*Tabernaemontana johnstonii*	*Dracaena afromontana*	*Symphonia globulifera*	*Maesa lanceolata*	*Xymalos monospora*	*Neoboutonia macrocalyx*	*Magnistipula butayei*	*Syzygium parvifolium*	Total (cm)²	Gi (m /ha²	%
[15-25[	0	0	86,2	187,3	0	48,8	57,5	313,	87,9	105,	0	0	0	0	20,	0	906,8	1,01	1,74
[25-35[	58	67	258,7	277,1	62,	0	0	107,	138,6	0	58	0	0	0	0	0	1027,4	1,14	1,97
[35-45[	0	229,9	274,2	103,2	0	0	0	0	280,9	0	0	121,	0	0	0	0	1009,3	1,12	1,93
[45-55[	207,1	0	199,1	0	0	0	0	0	0	0	232,2	0	0	223,	0	0	862	0,96	1,65
[55-65[	0	240,8	602,3	0	0	0	0	0	0	249,	0	0	0	0	0	0	1092,8	1,21	2,09
[65-75[	379,1	0	401,4	0	0	0	0	0	401,4	0	0	0	0	0	0	0	1181,9	1,31	2,26

[75-85[	561,8	459,9	0	1006,	0	0	0	0	472,1	0	0	0	459,	0	0	0	2960,1	3,29	5,67
[85-95[	0	1219,	0	644,9	0	0	0	0	630,7	0	0	0	0	0	0	0	2494,7	2,77	4,78
[95-105[	0	764,6	0	0	0	0	0	0	1755,	0	0	0	0	0	0	0	2520,2	2,80	4,83
[105-115[	0	0	0	894,6	0	0	0	0	911,5	0	0	0	0	0	0	0	1806,1	2,01	3,46
[115-125[	1165,7	0	1185	0	0	0	0	0	1052,	0	0	0	0	0	0	0	3403,6	3,78	6,52
[125-135[	0	1408,	0	0	0	0	0	0	1264	0	0	0	0	0	0	0	2672,4	2,97	5,12
[135-145[	0	1605,	0	0	0	0	0	0	0	0	0	0	0	0	0	1650,	3256,3	3,62	6,23
[155-165[	1987,6	1937,	0	0	0	0	0	0	0	0	0	0	0	0	0	0	3925,2	4,36	7,52
[165-175[	0	2167,	0	0	0	2410,	0	0	0	0	0	0	0	0	0	0	4578,1	5,09	8,77
[185-195[	0	2935	0	2784,	0	0	0	0	0	0	0	0	0	0	0	0	5719,2	6,35	10,9
[255-265[	0	0	5258,	0	0	0	0	0	0	0	0	0	0	0	0	0	5258,7	5,84	10,0
[305-315[	0	0	0	0	0	0	0	0	0	0	7552,	0	0	0	0	0	7552,9	8,39	14,4
Total (cm²)	4359,3	13035	8265,	5897,	62,	2459,	57,5	421	6995,	355	7843,	121,	459,	223,	20,	1650,	52228		
Gi (m /ha²)	4,84	14,48	9,18	6,55	0,0	2,73	0,06	0,47	7,77	0,39	8,71	0,13	0,51	0,25	0,0	1,83		58,03	
%	8,35	24,96	15,83	11,29	0,1	4,71	0,11	0,81	13,39	0,68	15,02	0,23	0,88	0,43	0,0	3,16			100

- **Litter quantification**

Figure 28 shows the variation in the number of elements making up the litter as a function of size. Smaller elements (level 1 and 2) are well represented, with around 71.65% of all litter collected, indicating good litter decomposition in the study area. Generally speaking, the number of litter items decreases with increasing size, with the exception of the Ndubura 2 site, where size 3 items outnumbered those of size 2. The Ndubura 2 and Ndubura 3 sites recorded a large amount of litter compared with the other sites, with 22.91% of all litter each. The absence of herbaceous cover favored litter collection at both sites. Larger items were collected at the Ndubura 1 and Ndubura 3 sites, where *Neoboutonia macrocalyx* and *Myrianthus holstii* with large leaves dominated, respectively.

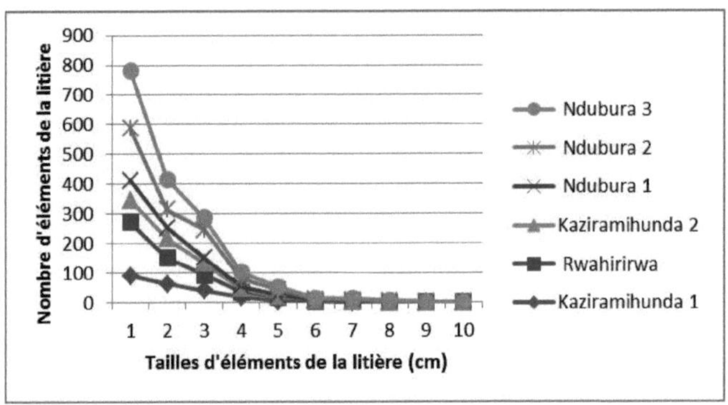

Fig. 28: Variation in litter element size by site

- **Comparison of basal area, wood density and number of trees from lalitière**

A comparison between basal area, wood density and
number of litter elements was carried out for different sites.

The Pearson correlation coefficient r calculated between basal area and wood density for the entire study area is low ($r = 0.4$). Generally, but not always, basal area is inversely proportional to tree density.

The higher the tree density, the lower the basal area. The lack of correlation is due to the Kaziramihunda 1 site, which deviates significantly from this theory.

Indeed, Kaziramihunda 1 has a very high tree density compared with the other sites, with a basal area that is not low. Thus, if we calculate r excluding Kaziramihunda 1, it is equal to 0.7.

The Pearson correlation coefficient r calculated between basal area and the number of litter elements with all data series is equal to 0.4, showing the absence of correlation between these two data series. Indeed, with the exception of Kaziramihunda 2, the other sites with a large basal area also have a high number of litter elements. Thus, r calculated after omitting Kaziramihunda 2 is equal to 0.8. This shows that, at Teza, large trees release large quantities of litter. At Kaziramihunda 2, it is the herbaceous cover that has generated the litter collection.

The low value of Pearson's correlation coefficient ($r = 0.1$) found between wood density and number of litter elements with all data series shows the absence of linearity. By calculating r while excluding the Kaziramihunda 1 site, it is equal to 0.7. At this site, the high wood density does not result in a higher number of litter elements than at other sites.

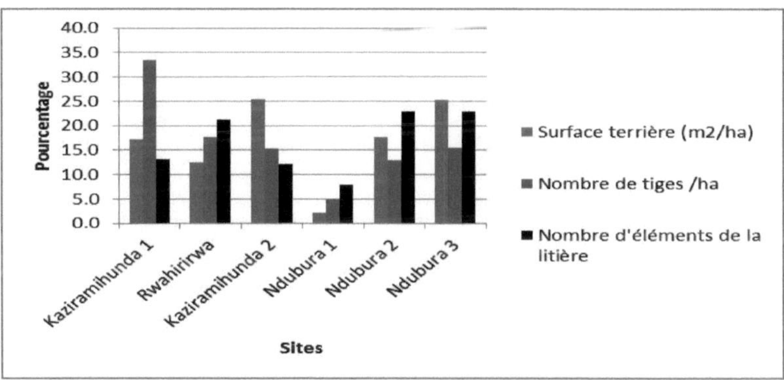

Fig. 29: Variation in basal area, wood density and number of trees elements of the Litter

IV.1.5. Habitat dynamics

- **Probable evolution of habitats in the study area**

- kaziramihunda 1 website

The Kaziramihunda 1 site is a secondary forest with *Macaranga kilimandscharica* and *Tabernaemontana johnstonii*, whose formula is A B S H_{1031++} G. Characteristic secondary forest species such as *Macaranga kilimandscharica, Tabernaemontana johnstonii, Polyscias fulva* and *Syzygium parvifolium* dominate the middle tree stratum. Overall, *Macaranga kilimandscharica* and *Tabernaemontana johnstonii* alone account for 48.2% of all stems measured and 37.4% of the basal area of the site. The same individuals appear in the lower tree stratum, with new species such as *Myrianthus holstii, Xymalos monospora, Bersama abyssinica, Strombosia scheffleri, Symphonia globulifera* and others. The shrub layer includes woody plants, mainly Rubiaceae such as *Galiniera coffeoides* and *Chassalia subochreata*. Individuals of the *Strombosia scheffleri* and *Symphonia globulifera* characteristic of primary forests also appear in the lower strata (herbaceous and shrub layer). The area is poor in herbaceous species, as the treetops prevent light from penetrating to the ground. The humic soil is covered with thick litter, forming piles in some places. No human activity has been observed on this site.

When the forest remains as it is, untouched by human or natural constraints, the presence of *Strombosia scheffleri* and *Symphonia globulifera* species characteristic of primary forests, especially in the lower strata, suggests that it could evolve into a *Strombosia scheffleri* and *Symphonia globulifera* primary forest in 10 years' time.

On the other hand, when the forest is subjected to anthropogenic actions (firewood harvesting, timber felling, etc.), the middle tree stratum may be eliminated and the forest may become a degraded secondary forest with *Macaranga kilimandscharica* and *Tabernaemontana johnstonii*. With the intensification of these actions, the forest could become a pre-forest recrû in 35 years' time.

Natural factors may also cause the forest to regress, depending on their intensity.

- *Rwahirirwa site*

This is a degraded secondary forest of *Syzygium parvifolium* with the formula A B S H$_{5144+}$ G. The middle tree stratum contains a single *Syzygium parvifolium*. Syzygium parvifolium also appears in the lower arborescent stratum, along with other species such as *Symphonia globulifera, Galiniera coffeoides, Psychotria palustris, Xymalos monospora, Macaranga kilimandscharica, Dracaena afromontana, Coccinia mildbraedii* and others. Creepers, mainly *Schefflera goetzenii, Schefflera abyssinica, Embelia Schimperi* and *Embelia libeniana*, are also found in this stratum, along with many epiphytes. Stems of *Macaranga kilimandscharica* are common in the lower tree stratum. Overall, *Syzygium parvifolium* alone accounts for 39.56% of basal area, but with no more stems than *Macaranga kilimandscharica*. The shrub layer mainly comprises *Psychotria palustris, Alchornea hirtella, Virectaria major, Allophylus chaunostachys, Chassalia subochreata, Embelia Schimperi, Schefflera abyssinica, Dracaena afromontana* and *Symphonia globulifera*. Ferns of the Aspleniaceae family, notably *Asplenium aethiopicum* and *Asplenium elliotii*, dominate the sub-shrub and/or herbaceous stratum, a sign of degradation. The gently sloping soil is covered with litter. Rwahirirwa is located close to a track joining the communes of Bukeye and Rugazi, which facilitates the cutting of dead wood at the site.

If the cutting of dead wood observed in Rwahirirwa is halted and natural factors do not become severe, individuals of *Symphonia globulifera*, a species characteristic of primary forests, will be able to develop and take the lead in the future. Under these conditions, Rwahirirwa could evolve into a *Symphonia globulifera* primary forest in 10 years' time.

If logging of dead wood continues, even the living trees may be removed, in which case the elimination of the middle tree layer will transform Rwahirirwa first into degraded secondary forest with *Macaranga kilimandscharica*, and then into pre-forest scrub in 35 years' time.

- *Kaziramihunda 2 site*

The Kaziramihunda 2 site is a primary forest with *Prunus africana*. The upper tree stratum includes giant *Prunus africana* trees, a species characteristic of primary forests, and species characteristic of secondary forests, notably *Polyscias fulva* and *Syzygium parvifolium*. The same species recur in the middle tree stratum, along with *Tabernaemontana johnstonii*, *Myrianthus holstii* and lianas such as *Schefflera goetzenii*, *Urera hypselodendron* and *Mimosa montana*. Dominant species in the lower tree layer include *Myrianthus holstii*, *Tabernaemontana johnstonii*, *Xymalos monospora*, *Dracaena afromontana*, *Syzygium parvifolium*, *Xymalos monospora* and *Magnistipula butayei*. The shrub layer is diverse. It is dominated by *Piper capense, Chassalia subochreata, Symphonia globulifera, Galiniera coffeoïdes, Strombosia scheffleri, Dracaena afromontana, Myrianthus holstii, Psychotria palustris* and others.

The treetops allow light to penetrate to the ground, encouraging the development of an extensive herbaceous cover consisting mainly of *Chassalia subochreata, Rourea thomsonii, Gynura scandens, Myrianthus holstii, Sericostachys scandens, Urera hypselodendron, Symphonia globulifera, Galiniera coffeoïdes, Setaria megaphylla* and many ferns, notably *Asplenium dregeanum, Asplenium elliotii* and *Asplenium glamosum*. The grasses are planted in unevenly distributed litter on deep, humic soil. Among the stems measured, *Prunus africana* had only 3 stems, but with large circumferences, giving the highest basal area value. The species *Myrianthus holstii* and *Tabernaemontana johnstonii* make up the majority of measured stems, but the sum of their basal areas is equal to that of the 3 *Prunus africana* stems mentioned above.

No human action was observed at Kaziramihunda.

2. The formula for this habitat is A B S H_{7321+} G.

Assuming that Kaziramihunda 2 remains protected from human and natural constraints, it will be able to evolve towards climax (the ultimate stage in the evolution of a forest). In the event of regressive evolution, Kaziramihunda 2 could become a secondary forest with *Myrianthus holstii* and *Tabernaemontana johnstonii* in 10 years' time.

- Ndubura 1 site

This is a pre-forest scrubland with the formula A B S H_{11910} G. With a few stems of *Neoboutonia macrocalyx, Dombeya goetzenii* and *Discopodium penninervium,* it's an airy place, dominated by the herbaceous stratum that completely covers the ground. There are three strata at Ndubura 1. The middle tree stratum comprises just two species, *Neoboutoniamacrocalyx* and *Dombeya goetzenii,* with 17 and 3 stems respectively out of 31 stems counted.

The shrub layer is dominated by *Brillantaisia cicatricosa, Gynura scandens, Rubus pinnatus, Discopodium penninervium, Triumfetta cordifolia, Neoboutoniamacrocalyx* and *Ensete ventricosum.* The sub-shrubby and/or herbaceous stratum is mainly represented by *Anisosepalum humbertii, Solanum nigrum, Urera hypselodendron, Piper capense, Brillantaisia cicatricosa, Gynura scandens, Rubus pinnatus* and typical wasteland species such as *Pteridium aquilinum* and *Clerodendrum johnstonii.* The loose, humic, deep soil is completely covered with grasses dominated by *Brillantaisia cicatricosa.* No anthropogenic action has been observed on this site.

If it evolves gradually, Ndubura 1 will become a dense forest in 35 years. Otherwise, it will fall to the pioneer stage in 5 years.

- Ndubura 2 site

This is a primary *Chrysophyllum gorungosanum* forest with the formula A B S H_{9711+} G. The upper tree stratum is dominated by giant trees of *Chrysophyllum gorungosanum* and *Strombosia scheffleri,* species characteristic of primary forests. This stratum also includes 3 stems of *Synsepalum attenuatum.* The medium tree stratum includes *Macaranga kilimandscharica, Chrysophyllum gorungosanum, Myrianthus holstii, Tabernaemontana johnstonii, Strombosia scheffleri* and *Polyscias fulva.* The aforementioned species reappear in the lower tree stratum, with the addition of *Xymalos monospora, Symphonia globulifera* and *Alchornea hirtella.* The treetops of the three preceding strata prevent light from penetrating to the ground, and the abundant litter deposits considerably impoverish the sub-shrub and/or herbaceous strata, thus justifying the site's specific poverty.

The shrub layer is dominated by *Strombosia scheffleri, Chassalia subochreata, Alchornea hirtella, Macaranga kilimandscharica, Dracaena afromontana, Rauvolfia mannii* and others. The sub-shrub and/or herbaceous stratum is less diverse. It includes *Alchornea hirtella, Sericostachys scandens, Gynura scandens, Myrianthus holstii, Chrysophyllum gorungosanum, Chassalia subochreata*, *Asplenium elliotii*, *Asplenium mannii*, *Dracaena afromontana* and *Galiniera coffeoides*. The humic soil is gently sloping with a lot of litter. Out of 108 stems measured, *Chrysophyllum gorungosanum* has only 6 stems, but is the first species to have a large basal area (40.07% of the total basal area). *Macaranga kilimandscharica* has 22 stems and 10.94% of the total basal area. No anthropogenic action was observed at Ndubura 2.

Assuming that Ndubura 2 remains undisturbed by human intervention, it will be able to evolve towards climax. Human intervention could transform Ndubura 2 into a secondary *Macaranga kilimandscharica* forest in 10 years' time. With the intensification of human activities, Kaziramihunda 2 could become an herbaceous wasteland in 40 years' time.

- *Ndubura 3 site*

Ndubura 3 is a primary *Chrysophyllum gorungosanum* forest with the formula A B S H$_{96210}$ G. The upper tree stratum is dominated by *Symphonia globulifera*, *Strombosia scheffleri* and *Chrysophyllum gorungosanum*, species characteristic of primary forests. The middle tree stratum includes *Macaranga kilimandscharica*, *Chrysophyllum gorungosanum*, *Strombosia scheffleri*, *Tabernaemontana johnstonii*, *Xymalos monospora*, *Myrianthus holstii* and *Symphonia globulifera*. The specific poverty of the lower strata is caused by the aforementioned tree crowns, which do not allow light to penetrate to the ground.

The lower tree layer is dominated by *Macaranga kilimandscharica*, *Neoboutonia macrocalyx*, *Tabernaemontana johnstonii*, *Chrysophyllum gorungosanum*, *Myrianthus holstii*, *Symphonia globulifera* and *Xymalos monospora*. The shrub layer is undiversified and includes *Chrysophyllum gorungosanum*, *Myrianthus holstii*, *Symphonia globulifera*, *Strombosia scheffleri*, *Chassalia subochreata* and *Xymalos monospora*. The sub-shrub and/or herbaceous stratum is species-poor. It mainly comprises *Mimulopsis solmsii*, *Asplenium dregeanum*, *Asplenium elliotii*, *Asplenium glamosum*, *Hypoestes forskalei*, *Sericostachys scandens*, *Chassalia subochreata*, *Rauvolfia manni*, *Strombosia scheffleri* and *Tabernaemontana johnstonii*. The soil is covered with heavy litter on a relatively gentle slope. Out of 97 stems measured, *Chrysophyllum gorungosanum* accounted for 13 stems, the first to have a large basal area, with 25% of the total basal area. *Strombosia scheffleri* is the first species with many stems, with 16 stems, but its basal area is negligible. *Myrianthus holstii* has 15 stems and is the second species with a high basal area, with 15.8% of the total basal area. No human action was observed at Ndubura 3.

With gradual evolution, Ndubura 3 will become primary forest having reached climax. If Ndubura 3 is disturbed by human or natural action, the primary forest will give way to a secondary forest of *Myrianthus holstii* in 10 years' time.

- **Guidelines for monitoring the dynamics of surveyed habitats**

- *Kaziramihunda 1 site*

The LEM data sheet (Table 29) highlights some of the elements to be taken as a reference for monitoring the dynamics of the Kaziramihunda 1 site. Species characteristic of each stratum have been reported. The two tree strata are dominated by *Polyscias fulva, Macaranga kilimandscharica, Tabernaemontana johnstonii, Myrianthus holstii* and *Syzygium parvifolium*. The shrub layer mainly comprises *Dracaena afromontana, Chassalia subochreata, Alchornea hirtella, Lindackeria kivuensis* and *Rauvolfia mannii*. The sub-shrub and/or herbaceous stratum is dominated by *Sericostachys scandens* and many ferns of the Apleniaceae family, notably *Asplenium elliotii* and *Asplenium sandersonii*. No anthropogenic action has been observed at this site. No signs of fauna were observed.

Table 29: Habitat dynamics monitoring sheet for the Teza sector of KNP (Kaziramihunda 1 site)

GPS	Alt: 2128 m		S 03°12.771'		E 029°32.810			Topo: 50% off	Date: May 01, 2013
Human activities			**Fauna**		**Habitat observations**				
Type	Surf	Sign	Species	Nbr	Type	Strata	Ht	Rc	Species
						A-TGA			
					Secondary forest with *Macaranga Kilimanscharica* and *Tabernaemontana johnstonii*	A-GA	25 m	70%	*Polyscias fulva;* *Sapium ellipticum;* *Syzygium parvifolium;* *Macaranga kilimandscharica;* *Tabernaemontana johnstonii*
						A-AM	13 m	80%	*Tabernaemontana johnstonii;* *Myrianthus holstii;* *Macaranga kilimandscharica;* *Strombosia scheffleri;* *Symphonia globulifera*
						aB	5 m	30%	*Alchornea hirtella;* *Lindackeria kivuensis;* *Rauvolfia mannii;* *Dracaena afromontana;* *Chassalia subochreata*
						SsAH	0,7 m	10%	*Sericostachys scandens;* *Asplenium elliotii;* *Asplenium sandersonii;* *Kalanchoe crenata;* *Impatiens burtonii*

Formula: A10B3S1H+G+

Legend (explanation of abbreviations)

Observation reference	Human activities	Fauna	Habitat observations	
			Strata	**Height (m)**
Lat: latitude	**Surf**: Surface	**Nbr**: Number	**A-TGA**: Arborescent with very large trees	30< 50
Lon: Longitude			**A-GA**: Arborescent with large trees	20<30
			A-AM: Arborescent with small to medium-sized trees	7<20
			aB: shrubby	2<7
			SsAH: Sub-shrubby and/or herbaceous	<2
Topo: Topography			**Ht**: Height (m)	
			Rc: Recovery (%)	

- *Rwahirirwa site*

The LEM data sheet (Table 30) highlights certain elements to be taken as a reference for monitoring the dynamics of the Rwahirirwa site. The dominant species in each stratum have been identified. The two tree strata are dominated by *Syzygium parvifolium*, *Tabernaemontana johnstonii* and *Macaranga kilimandscharica*. The shrub layer mainly comprises *Dracaena afromontana*, *Chassalia subochreata*, *Galiniera coffeoides*, *Psychotria palustris* and others. *Schefflera goetzenii* is a characteristic liana that tends to occupy all strata. The sub-shrubby and/or herbaceous stratum is dominated by *Asplenium aethiopicum*, *Asplenium elliotii*, *Begonia meyeri-johannis* and *Virectaria major*. A 2 m cut of dead wood[2] and a chimpanzee nest were found at this site.

Table 30: Habitat dynamics monitoring sheet for the Teza sector of KNP (Rwahirirwa site)

GPS		Alt: 2192 m		S 03°12.798'		E 029°32.698'			Topo: 45	Date: Le 04/05/2013
Human activities			Fauna			Habitat observations				
Type	Surf	Sign	Species	Nbr	Type	Strata	Ht	Rc		Species
Cut	2m²	1 nest	Chimpanzee	1	Degraded secondary forest with *Syzygium parvifolium*	A-TGA	35 m	20%		- *Syzygium parvifolium*
						A-GA				
						A-AM	14 m	50%		- *Tabernaemontana johnstonii;* - *Xymalos monospora;* - *Syzygium parvifolium;* - *Macaranga kilimandscharica;* - *Schefflera goetzenii*
						aB	5 m	15 %		- *Psychotria palustris;* - *Chassalia subochreata;* - *Galiniera coffeoides;* - *Dracaena afromontana* - *Mimulopsis solmsii*
						SsAH	0,5 m	80%		- *Asplenium aethiopicum;* - *Asplenium elliotii;* - *Begonia meyeri-johannis;* - *Virectaria major* - *Impatiens burtonii*

Formula: A5B1S4H4G+

Legend (explanation of abbreviations)

Observation reference	Human activities	Fauna	Habitat observations	
			Strata	**Height (m)**
Lat: latitude	**Surf**: Surface	**Nbr**: Number	**A-TGA:** Arborescent with very large trees	30< 50
Lon: Longitude			**A-GA:** Arborescent with large trees	20<30
			A-AM: Arborescent with small to medium-sized trees	7<20
			aB: shrubby	2<7
			SsAH: Sub-shrubby and/or herbaceous	<2
Topo: Topography			**Ht**: Height (m)	
			Rc: Recovery (%)	

- *Kaziramihunda 2 site*

This is the site with the highest species richness compared to the others. Species such as *Prunus Africana, Polyscias fulva, Syzygium parvifolium, Myrianthus holstii, Xymalos monospora, Dracaena afromontana* and *Tabernaemontana johnstonii* are characteristic of the three arborescent strata. The shrub layer is dominated by *Galiniera coffeoïdes, Dracaena afromontana, Piper capense, Strombosia scheffleri* and others. The herbaceous stratum mainly comprises *Setaria megaphylla, Sericostachys scandens, Asplenium dregeanum, Asplenium elliotii, Asplenium glamosum*, etc. No human actions or animal tracks have been observed at this site. The LEM data sheet (Table 31) highlights a number of elements to be taken as a reference for monitoring the dynamics of the Kaziramihunda 2 site.

Table 31: Habitat dynamics monitoring sheet for the Teza sector of KNP (Kaziramihunda 2 site)

GPS	Alt: 2169 m		S 03°12.820'		E 029°32.653'				Topo: 45	Date: 06/05/2013
Human activities			Fauna		Habitat observations					
Type	Surf	Sign	Species	Nbr	Type	Strata	Ht	Rc		Species
						A-TGA	35 m	40%		*Prunus Africana;* *Polyscias fulva;* *Syzygium parvifolium*
						A-GA	25 m	50%		*Prunus africana;* *Polyscias fulva;* *Myrianthus holstii;* *Tabernaemontana johnstonii;* *Syzygium parvifolium*
						A-AM	13m	25%		*Myrianthus holstii;* *Tabernaemontana johnstonii;* *Prunus Africana;* *Dracaena afromontana;* *Magnistipula butayei;*
					Primary forest with *Prunus africana*	aB	4 m	25%		*Piper capense;* *Dracaena afromontana;* *Symphonia globulifera;* *Galiniera coffeoïdes;* *Strombosia scheffleri*
						SsAH	0,8 m	50%		*Setaria megaphylla;* *Sericostachys scandens;* *Asplenium dregeanum;* *Asplenium elliotii;* *Asplenium glamosum*

Formula: A7B3S2H1G+

Legend (explanation of abbreviations)

Observation reference	Human activities	Fauna	Habitat observations	
			Strata	**Height (m)**
Lat: latitude	**Surf**: Surface	**Nbr**: Number	**A-TGA:** Arborescent with very large trees	30< 50
Lon: Longitude			**A-GA:** Arborescent with large trees	20<30
			A-AM: Arborescent with small to medium-sized trees	7<20
			aB: shrubby	2<7
			SsAH: Sub-shrubby and/or herbaceous	<2
Topo: Topography			**Ht**: Height (m)	
			Rc: Recovery (%)	

- Ndubura 1 site

The LEM sheet (Table 32) shows the observations for monitoring the dynamics of the Ndubura 1 site. With a few individuals of *Neoboutonia macrocalyx* reaching an average height of 15 m, this site is dominated by the herbaceous stratum represented mainly by *Brillantaisia cicatricosa, Gynura scandens, Pteridium aquilinum, Solanum nigrum* and *Urera hypselodendron*. This pre-forest scree is the least evolved habitat in the entire study area, with no human action or signs of animal presence.

Table 32: Habitat dynamics monitoring sheet for the Teza sector of KNP (Ndubura 1 site)

GPS	Alt: 2216 m		S 03°12.908'		E 029°32.414'			Topo: 70% off	Date: 06/06/2013
Human activities			**Fauna**		**Habitat observations**				
Type	Surf	Sign	Species	Nbr	Type	Strata	Ht	Rc	Species
						A-TGA			
						A-GA			
						A-AM	15	1%	Neoboutonia macrocalyx; Dombeya goetzenii
					Pre-forest harvesting	aB	3 m	5%	Brillantaisia cicatricosa; Gynura scandens; Discopodium penninervium; Neoboutonia macrocalyx; Ensete ventricosum
						SsAH	1,5 m	100%	Solanum nigrum; Urera hypselodendron; Brillantaisia cicatricosa; Gynura scandens; Pteridium aquilinum

Formula: A1B1S9H1G0

Legend (explanation of abbreviations)

Observation reference	Human activities	Fauna	Habitat observations	
Lat: latitude	Surf: Surface	Nbr: Number	**Strata**	**Height (m)**
Lon: Longitude			A-TGA: Arborescent with very large trees	30< 50
			A-GA: Arborescent with large trees	20<30
			A-AM: Arborescent with small to medium-sized trees	7<20
			aB: shrubby	2<7
Topo: Topography			SsAH: Sub-shrubby and/or herbaceous	<2
			Ht: Height (m)	
			Rc: Recovery (%)	

- Ndubura 2 site

The LEM sheet (Table 33) shows the observations made to monitor the dynamics of the Ndubura 2 site. Species such as *Chrysophyllum gorungosanum, Synsepalum attenuatum, Strombosia scheffleri, Polyscias fulva, Myrianthus holstii, Tabernaemontana johnstonii, Xymalos monospora, Symphonia globulifera, Macaranga kilimandscharica*, etc, characterize the three tree strata. The shrub layer is dominated by *Dracaena afromontana, Rauvolfia mannii, Chassalia subochreata* and *Alchornea hirtella*. The herbaceous layer is dominated by species such as *Sericostachys scandens, Gynura scandens* and many ferns of the Aspleniaceae family, notably *Asplenium eiliotii* and *Asplenium manii*. This habitat has not been subject to any anthropogenic action or signs of fauna.

Table 33: Habitat dynamics monitoring sheet for the Teza sector of KNP (Ndubura 2 site)

GPS	Alt: 2260 m			S 03°12.913'		E 029°32.380'				Topo: 40% of	Date: 08/06/ 2013
Human activities			**Fauna**			**Habitats observation**					
Type	Surf	Sign	Species	Nbr	Type	Strata	Ht	Rc			Species
					Chrysophyllum gorungosanum primary forest	A-TGA	35 m	50%			- *Chrysophyllum gorungosanum;* - *Synsepalum attenuatum;* - *Strombosia scheffleri*
						A-GA	25 m	30%			- *Macaranga kilimandscharica;* - *Chrysophyllum gorungosanum;* - *Polyscias fulva;* - *Myrianthus holstii;* - *Tabernaemontana johnstonii*
						A-AM	12 m	30%			- *Myrianthus holstii;* - *Xymalos monospora;* - *Symphonia globulifera* - *Chrysophyllum gorungosanum;* - *Strombosia scheffleri;*
						aB	4 m	15%			- *Chassalia subochreata;* - *Alchornea hirtella;* - *Macaranga kilimandscharica;* - *Dracaena afromontana;* - *Rauvolfia mannii*
						SsAH	0,5 m	15%			- *Sericostachys scandens;* - *Gynura scandens;* - *Galiniera coffeoides;* - *Asplenium elliotii;* - *Asplenium manii;*

Formula:A9B7S1H1G+

Legend (explanation of abbreviations)

Observation reference	Human activities	Fauna	Habitat observations	
			Strata	**Height (m)**
Lat: latitude	**Surf**: Surface	**Nbr**: Number	**A-TGA:** Arborescent with very large trees	30< 50
Lon: Longitude			**A-GA**: Arborescent with large trees	20<30
			A-AM: Arborescent with small to medium-sized trees	7<20
			aB: shrubby	2<7
Topo: Topography			**SsAH**: Sub-shrubby and/or herbaceous	<2
			Ht: Height (m)	
			Rc: Recovery (%)	

- Ndubura 3 site

The LEM sheet (Table 34) shows the observations for monitoring the dynamics of the Ndubura 3 site. Species such as *Symphonia globulifera, Chrysophyllum gorungosanum, Strombosia scheffleri, Myrianthus holstii, Tabernaemontana johnstonii, Neoboutonia macrocalyx, Macaranga kilimandscharica, Xymalos monospora*, etc, characterize the three arborescent strata. The shrub layer is dominated by *Chassalia subochreata, Xymalos monospora, Myrianthus holstii, Symphonia globulifera* and others. The herbaceous cover is dominated by ferns of the Aspleniaceae family, interspersed with young shoots of the above-mentioned species. This habitat has not been subject to any anthropogenic action or signs of fauna.

Table 34: Habitat dynamics monitoring sheet for the Teza sector of KNP (Ndubura 3 site)

GPS	Alt: 2281 m		S 03°12.887'		E 029°32.312'			Topo: 30% off		Date: 10/06/2013
Human activities		**Fauna**			**Habitat observations**					
Type	Surf	Sign	Species	Nbr	Type	Strata	Ht	Rc		Species
					Chrysophyllum gorungosanum primary forest	A-TGA	35 m	65%		- *Symphonia globulifera;* - *Chrysophyllum gorungosanum;* - *Strombosia scheffleri*
						A-GA	25 m	50%		- *Macaranga kilimandscharica;* - *Chrysophyllum gorungosanum;* - *Tabernaemontana johnstonii;* - *Xymalos monospora;* - *Myrianthus holstii;*
						A-AM	15 m	20%		- *Tabernaemontana johnstonii;* - *Neoboutonia macrocalyx;* - *Myrianthus holstii;* - *Symphonia globulifera;* - *Xymalos monospora*
						aB	5 m	60%		- *Chrysophyllum gorungosanum;* - *Myrianthus holstii;* - *Chassalia subochreata* - *Strombosia scheffleri;* - *Xymalos monospora*
						SsAH	0,8 m	30%		- *Asplenium dregeanum;* - *Asplenium elliotii;* - *Hypoestes forskaolii;* - *Chassalia subochreata ;* - *Rauvolfia mannii*

Formula:A9B6S2H1G0

Legend (explanation of abbreviations)

Observation reference	Human activities	Fauna	Habitat observations	
Lat: latitude	**Surf**: Surface	**Nbr**: Number	**Strata**	**Height (m)**
Lon: Longitude			**A-TGA:** Arborescent with very large trees	30< 50
			A-GA: Arborescent with large trees	20<30
			A-AM: Arborescent with small to medium-sized trees	7<20
			aB: shrubby	2<7
N: Number			**SsAH**: Sub-shrubby and/or herbaceous	<2
Topo: Topography			**Ht**: Height (m)	
			Rc: Recovery (%)	

IV.2. DISCUSSION

IV.2.1. Floristic characteristics

IV.2.1.1. Floristic richness

In our study area, considering systematic groups, dicotyledons dominate with 129 (77.71%) of all species inventoried. The most represented families are Rubiaceae, Asteraceae, Acanthaceae, Aspleniaceae, Orchidaceae, Euphorbiaceae, Urticaceae, Verbenaceae and Fabaceae. Manirakiza (2013), in Rwegura, found that the most represented families were: Asteraceae, Rubiaceae, Orchidaceae, Aspleniaceae, Euphorbiaceae, Urticaceae, Fabaceae, Acanthaceae and Apocynaceae. Wibereho (2010), in Bugarama (Kibira), found that the most represented families were: Asteraceae, Rubiaceae, Euphorbiaceae, Fabaceae, Solanaceae, Acanthaceae, Rosaceae, Urticaceae, Cucurbitaceae and Polygonaceae. Habiyaremye (1995), who worked in the forests (Gishwati, Mukura, Nyungwe) of the Congo-Nile ridge, compiled a list similar to the present one. The results are comparable.

In terms of plant groupings, 166 species in 132 genera and 73 families were inventoried. The Ndubura 2 site recorded fewer species. This could be explained by the virtual non-existence of the herbaceous stratum, due to tree crowns preventing light penetration to the ground. This lack of specificity may also be linked to the presence of *Urera hypselodendrum* and *Simirestis goetzenii*, which are lianas considered to be competitive and which can handicap the regeneration and development of the forest. The Kaziramihunda 2 site has more species. This could be explained by the good penetration of light to the ground, which favours the development of the herbaceous stratum. The difference in species numbers between the study sites could also be explained by their stage of evolution.

Indeed, the less evolved sites (Ndubura 1 and Rwahirirwa) and the more evolved sites (Ndubura 2 and Ndubura 3) have fewer species than the moderately evolved Kaziramihunda 1 site. The exception is the Kaziramihunda 2 site, which has more species despite being more advanced. This situation shows that floristic composition varies from site to site.

Sorensen's similarity index values show that most of the sites surveyed have significant floristic similarities, which is the case for Kaziramihunda 1 and Rwahirirwa (K = 60.7), Kaziramihunda 1 and Kaziramihunda 2 (K = 62.3), Ndubura 2 and Ndubura 3 (K = 56.4) and many others. This shows that these sites have many common species. Dendrogram analysis reveals a clear dissimilarity between sites. The value of the specific quotient calculated for different sampled sites varies from 1.06 to 1.14, indicating the maturity of their flora. Hakizimana et *al*, 2011, found specific quotients of 1.44 and 1.37 respectively for the Kigwena and Rumonge forests. These results show that Teza remains more stable than Kigwena and Rumonge.

IV.2.1.2. Biological forms

Analysis of biological forms revealed that phanerophytes dominate with 63.8% of the gross spectrum and 93.3% of the weighted spectrum in the study area, highlighting the forested character of Kibira (Teza). Manirakiza (2013) found that phanerophytes dominate over other biological forms with a gross spectrum of 59% at Rwegura, Wibereho (2010) found that phanerophytes dominate with a gross spectrum of 60.2% at Bugarama, Lewalle (1972) found a gross spectrum of 68.5% and 76% of the weighted spectrum at the level of the middle horizon, Habonayo and Ndihokubwayo (2011) found that phanerophytes accounted for 72% and 67.9% of the gross spectrum in the degraded and non-degraded zones respectively at Rwegura, while Karisabiye and Nduwimana (2001) found 71.9% of the phanerophyte gross spectrum at Mpotsa. The results of the present study do not differ from those of these authors.

The significant proportions of champhytes (15% of the gross spectrum and 2.2% of the weighted spectrum) are linked to the light stress tolerance strategy. A comparison of biological forms between sites shows that the most advanced sites (Ndubura 2, Ndubura 3 and Kaziramihunda 2) have high levels of phanerophytes. Habiyaremye (1995) points out that the importance of phanerophytes increases with the degree of maturity of the phytocenoses. However, at the Rwahirirwa site, the proportion of champhytes and therophytes increases at the expense of phanerophytes compared with other sites. This leads us to conclude that the sites sampled are at different stages of evolution.

IV.2.1.3. Phytogeographical elements

Analysis of the phytogeographical distribution of species collected in the Teza sector of KNP shows that montane species occupy first place with 42.7% of the gross spectrum. The high proportion of these species proves that this area is typically montane. Manirakiza (2013), who worked in the Rwegura sector of KNP, found that montane species dominate with 44.1% of the gross spectrum. Habiyaremye (1993), who worked on the forests (Gishwati, Mukura, Nyungwe) of the Zaire-Nile Ridge, found that the majority of species (45%) have an afromontane distribution for primary forests. These results place the various study areas in the afromontane domain at the level of the mountain rainforest district.

The rate of wide-ranging species found (8.9%) is lower than the rates found by Manirakiza (2013) (12.4%) at Rwegura, by Wibereho (2010) (24.7%) at Bugarama, by Karisabiye and Nduwimana (2001) (34.30%) at Mpotsa, a sign that Teza is less disturbed given that the high rate of these species is, according to Lewalle (1972), an index of disturbance or banality of plant formations. The Ndubura 1 and Rwahirirwa sites are rich in these species, proving that the sites are at different levels of evolution. Considering the montane base element retained, the omni-montane group (Mo) dominates throughout the study area.

IV.2.2. Structural features

Physiognomically, the vegetation in the Teza sector of KNP is divided into 5 strata: the upper tree stratum, the middle tree stratum, the lower tree stratum, the shrub stratum and the sub-shrub and/or herb stratum. Giant *Prunus africana* and *Polyscias fulva* trees were found in the middle horizon (Kaziramihunda 2). These results are not dissimilar to those of Lewalle (1972), who found giant trees of the same species in this horizon. The sites sampled are physiognomically different. Kaziramihunda 1 and Rwahirirwa are secondary mountain forests, Kaziramihunda 2, Ndubura 2 and Ndubura 3 are primary forests, while Ndubura 1 is a wasteland.

Considering the distribution of stems by circumference class, 55% of the stems measured had a circumference of between 15 and 34 cm. The high proportion of stems with small circumferences could indicate the existence of natural conditions favorable to natural regeneration in recent years for the species observed. These results are similar to those of Manirakiza (2013), who found 61% of stems with a circumference between 15 and 29 cm in Rwegura, but very different from those of Wibereho (2010), who found 80% of stems with a circumference between 10 and 29 cm in Bugarama. This difference indicates that the habitats surveyed are more advanced than those at Rwegura and Bugarama.

In terms of wood density, it is 641 feet/ha with a circumference exceeding 30 cm. This is higher than the 396 ft/ha found by Manirakiza (2013) in Rwegura, excluding bamboo, and the 330 ft/ha found by Wibereho in Bugarama. This would mean that Teza remains more or less dense compared with Rwegura and Bugarama. The analysis reveals that wood density differs according to site and plant type.

Looking at stems over 34 cm in circumference, the Kaziramihunda 1 site comes out on top with 891 plants/ha, while the Ndubura 1 site ranks last with 113 plants/ha, reinforcing the idea that the sites are evolving in different ways.

The calculated basal areas classify Kaziramihunda 1, Rwahirirwa and Ndubura 2 as dense rainforests according to the Mosango and Lejoly (1990) classification, which considers the range between 23 and 50 m^2/ha for this forest type. Kaziramihunda 2 and Ndubura 3 exceed this limit, while the low basal area (5.2 m^2 /ha) of Ndubura 1 classifies it as wasteland. This confirms that the sites are at different stages of evolution.

Litter quantification shows that the percentage of small-sized elements (level 1 and 2) is high with around 71.65% of all litter collected, highlighting good litter decomposition in our study area. Manirakiza (2013) made the same constant at Rwegura level where he collected a large quantity of level 1 and 2 elements with 70.87% of all litter collected. Nzigidahera (2012) was able to show that litter is thinner and therefore in the process of decomposing in mountain forests than in lowland forests, but there is no shortage of exceptions.

The Pearson correlation coefficient r calculated between basal area and wood density for the entire study area is low ($r = 0.4$), but after omitting Kaziramihunda 1 we have a significant correlation ($r = 0.7$), which is due to the fact that this site records a high wood density with a basal area that is not low compared to the other sites.

The Pearson correlation coefficient r calculated between basal area and the number of litter elements after omitting Kaziramihunda 2 is significant (r = 0.8), inferring that at Teza level, these are large trees that release a large amount of litter. These results are not dissimilar to those of Manirakiza (2013), who found r = 0.80 from Rutongati 2. The correlation coefficient between wood density and amount of litter calculated using all data sets is low (r = 0.1).

IV.2.3. Habitat dynamics

The habitats surveyed are at different stages of evolution. Some are at an advanced stage of evolution, such as Kaziramihunda 2, Ndubura 2 and Ndubura 3, which are primary forests. Others are not very advanced, such as Kaziramihunda 1 and Rwahirirwa, which are secondary forests. Ndubura 1 is the least evolved site, constituting a wasteland belonging to the pre-forest recrûs.

According to the phytodynamic stages coexisting on the Congo-Nile ridge established by Habiyaremye (1995), in the event of progressive evolution, Kaziramihunda 1 will become a primary forest with *Strombosia scheffleri* and *Symphonia globulifera* in 10 years, Rwahirirwa will become a primary forest with *Symphonia globulifera* in 10 years, Kaziramihunda 2 will evolve towards the climax, Ndubura 1 will become a dense forest in 35 years, Ndubura 2 and Ndubura 3 will evolve towards the climax. In the event of regressive evolution, Kaziramihunda 1 will become a pre-forest recrû in 35 years, Rwahirirwa will become a wasteland belonging to the pre-forest recrû in 35 years, Kaziramihunda 2 will become secondary forest with *Myrianthys holstii* and *Tabernaemontana johnstonii* in 10 years, Ndubura 1 will fall to the pioneer stage in 5 years, Ndubura 2 will become herbaceous wasteland in 40 years, Ndubura 3 will become secondary forest with *Myrianthys holstii* in 10 years.

The habitats surveyed showed no signs of regressive evolution. With the exception of the Rwahirirwa site, where a cut of dead wood was noted, no other human action was observed in our study area. Given this situation, it is to be hoped that human actions will remain minimal, which will encourage the gradual evolution of the study area.

CONCLUSION AND RECOMMENDATIONS

The overall aim of our work, entitled **"Etude de la végétation du Parc National de la Kibira: *Secteur Teza*"** ("Vegetation study of the Kibira National Park: *Teza sector*"), was to build up the database needed to monitor the dynamics of habitats, populations and species in KNP management. The initial hypotheses were that KNP habitats are at different levels of evolution, that floristic and structural characteristics vary from one habitat to another, and that KNP habitats are following a regressive evolution. Six sites - Kaziramihunda 1, Rwahirirwa, Kaziramihunda 2, Ndubura 1, Ndubura 2 and Ndubura 3 - were defined along a single transect.

In terms of floristic composition, 166 species in 132 genera and 73 families were inventoried. Dicotyledons dominate with 129 species, i.e. 77.71% of all species. The Rubiaceae (8.43%), Asteraceae (4.82%), Aspleniaceae (4.82%), Acanthaceae (4.22%) and orchidaceae (4.22%) families are the most dominant. The Ndubura 2 site is the least phytodiverse with 50 species, i.e. 30.12% of all species inventoried, while the Kaziramihunda 2 site is the most phytodiverse with 104 species, i.e. 62.65% of all species inventoried. Sorensen's similarity index valucs found by comparing different sites show that most of them have many species in common, while the dendrogram shows a clear dissimilarity. The specific quotient values for the different sites, ranging from 1.06 to 1.14, testify to the maturity of their flora.

Analysis of the biological forms shows the dominance of phanerophytes (63.8% of the raw spectrum), followed by chamephytes (15% of the raw spectrum). The high proportion of phanerophytes shows that the forest character has been preserved at Teza, while the significant proportion of chamephytes reflects a strategy of adaptation to light stress. An analysis of the phytogeographical elements reveals the predominance of montane species with 42.7% of the gross spectrum, placing the study area in the afromontane domain at the level of the mountain rainforest district.

In terms of stand density, 831 stems were counted, of which 134, or 16.1%, were *Macaranga kilimandscharica*. 55% had circumferences between [15-35[cm. The large number of stems with small circumferences indicates good forest regeneration in the study area. Wood density throughout the study area was estimated at 1187 plants/ha. The surface areas of the Kaziramihunda 1,

Rwahirirwa and Ndubura 2 sites classify them as dense rainforests in the sense of Mosango and Lejoly (1990), while Kaziramihunda 2 and Ndubura 3 exceed the classification limits. The Ndubura 1 site has only 5.2 m^2 /ha and is classified as wasteland. Physiognomically, strata ranging from 3 to 5 were observed throughout the study area. Litter decomposition is very high throughout the study area. The Kaziramihunda 2 site had a small amount of litter compared with other, more advanced sites. The basal area-wood density relationship for all data series shows a non-significant correlation (r = 0.4), but after omitting the Kaziramihunda 1 site, we have a good correlation (r = 0.7).

Basal area and litterfall for all sites do not show good linearity (r = 0.4), but after omitting the Kaziramihunda 2 site we have a good correlation (r = 0.8). The ligneous density-litter quantity relationship for the entire study area shows poor linearity (r = 0.1).

The habitats at the surveyed sites are at different stages of evolution. The Kaziramihunda 2, Ndubura 2 and Ndubura 3 sites are primary forests, while the Kaziramihunda 1 and Rwahirirwa sites are secondary forests. The least evolved Ndubura 1 site is classified as pre-forest scrub. These sites will evolve differently over time. With the exception of the Rwahirirwa site, where a cut of dead wood was observed, no other anthropogenic action was observed in the study area. There was a total absence of tree stumps that might have been felled. Hence, as we have just shown, one of our hypotheses that KNP habitats are following a regressive evolution is invalidated. On the contrary, the majority of the sites surveyed show an upward trend.

Considering that deforestation and the loss of genetic diversity are some of the issues currently at the forefront of international public concern, aware of the real danger posed by the unprecedented disappearance of KNP's rich flora and fauna, and convinced that our heritage has not yet completely lost its value, the following recommendations have been formulated:

-    Enforce the various laws that exist in the field to protect
protected areas in Burundi in general and the KNP in particular;
-    Increase the number of forest rangers to cover the entire protected area and provide them with equipment better suited to their tasks (boots, raincoats, communication tools, etc.);
-    Use durable materials to mark our transect, given that further research can

be carried out along it to enhance the results of the present study;

-   Multiply studies on the KNP to get to know this forest massif better, especially by making an inventory of all its flora;

-   Plan similar studies on the same sites surveyed, especially in the years to come, with the aim of rigorously monitoring the vegetation dynamics of KNP.

BIBLIOGRAPHY

Bangirinama, F., (2010). Ecosystemic restoration processes during post-cultivation dynamics in Burundi: Mechanisms, characterization and ecological series. Doctoral thesis, Fac. Sc. ULB, 222 p.

Cordonnier, T., (2014). Basal area: measurement methods and interests, 8p.

FAO, (2002). Case study of exemplary forest management in Central Africa: Kibira National Park, Burundi. Working paper FM/9F. Forest Resources Development Service, Forest Resources Division. FAO, Rome, 32 p.

Fischer, E., Killmann, D., Delepierre, G. & Lebel J.-P., (2010).The orchids of Rwanda.Koblenz Geographical Colloquia, Series Biogeographical Monographs 2, 439 p.

Fischer, E., Killmann, D., (2008). Illustrated Field Guide to the plants of Nyungwe National Park Rwanda, Koblenz Geographical Colloquia.Series Biogeographical Monographs1, 771 p.

Gourlet, S., (1986). Le Parc National de la Kibira au Burundi ; quelles potentialités pour quel avenir ? Rapport de stage, ENGREF (Montpelliet), 97 p.

Habiyaremye, F.M., (2012). Suivi de la dynamique des habitats dans les aires protégées en République du Burundi, syllabus, 83 p.

Habiyaremye, F.M., (1995). Phytocoenological study of the eastern ridge of Lake Kivu (Rwanda). Doctoral thesis, Fac. Sc. ULB, 371 p.

Habiyaremye, F.M., (1993). Analyses phytosociologique des forêts primaires mésophiles de la Crête Zaïre- Nil au Rwanda; *Belgian journal of Botany*, n° 126 (1): 100- 135.

Habonayo, R., Ndihokubwayo, N., (2011). Determination of degradation indicators in Kibira National Park (Burundi): Cas du secteur Rwegura. Mémoire de Master complémentaire en Sciences de l'environnement, 61p.

Habonimana, B., Nzigidahera, B., Cimanimpaye, C., (2007). Etude d'exploitation et de conservation d'*Arundinalia alpina* Michaux, espèce menacée d'extinction au Burundi, *Bulletin Scientifique de l'I.N.E.C.N* 4: 3-8.

Habonimana, B., Nzigidahera, B., Inamahoro M., (2007). Approche participative d'identification des espèces végétales autochtones menacées au Burundi: Diagnostic des connaissances traditionnelles, *Bulletin Scientifique de l'I.N.E.C.N* 2: 10-16.

Hakizimana, P., (2012). Analysis of the composition, spatial structure and natural plant resources collected in the Kigwena dense forest and Rumonge clear forest in Burundi. PhD thesis, Université Libre de Bruxelles, 247 p.

Hakizimana, P., Bangirinama, F., Habonimana, B., Bogaert, J., (2011). Comparative analysis of the flora of the dense forest of Kigwena and the clear forest of Rumonge in Burundi, *Bulletin Scientifique de l'I.N.E.C.N* 9: 53-61.

IGEBU, (2015). Monthly climatological bulletin (1984- 2014).

Krug, O., (1993). Studies of production and agrarian systems in the three riparian communes of the Kibira National Park: proposals for conflict reduction. Mémoire de DSPV Formation Supérieure Tropicale du CIHEAM, INECN, 71 p.

Lewale, J., (1972). The vegetation stages of western Burundi.
Bulletin of the National Botanic Garden of Belgium, 42 (1/2): 1-247.
Manirakiza, M., (2013). Etablissement de la situation de référence dans le but du suivi de la dynamique des habitats au Parc National de la Kibira: cas du secteur Rwegura. Mémoire d'Ingénieur Agronome; UB (FACAGRO), 111p.

Masharabu, T., (2012). Flora and vegetation of the Ruvubu National Park in Burundi: diversity, structure and implications for conservation. PhD thesis, Université Libre de Bruxelles, 224 p.

Nderagakura, D., (1999). Etude de la dynamique des infractions et approche rationnelle de gestion communautaire des aires protégées du Burundi: cas du parc national de la Kibira. Mémoire d'Ingénieur Agronome; UB (FACAGRO), 80 p.

Ndayikeza, W., Niyimpaye, J., (2007). Contribution à l'étude des interrelations entre la population de Musigati et le Parc National de la Kibira. Mémoire d'Ingénieur Industriel, UB,69 p.

Niyukuri, J., (2012). Analysis of the forest edge effect in Kibira National Park: the case of anthropogenic edges in the Rwegura sector. Mémoire de Master complémentaire en Sciences de l'environnement, 92 p.

Ntibarirarana, R., (2002). Contribution à l'étude des ressources végétales exploitables du Parc National de la Kibira. Mémoire d'Ingénieur Industriel, UB, 75 p.

Nzigidahera,B., (2012). On the relationship between forest physiognomy and litter structure and abundance: Methodological approach and application to natural forests in Burundi. *Bulletin Scientifique de l'I.N.E.C.N* 10: 46-61.

Nzigidahera, B., (2000). Study of national biodiversity and identification of priorities for its conservation. Bujumbura- Burundi, 125 p.

Ramade, F., (2009). Eléments d'écologie. Ecologie fondamentale, 4ème
edition, Paris, 656 p.
Salvaudon, A., (2006). Gestion des milieux et des espèces, Memento de terrain n° 83, 3p.

Sedjar, A., (2012). Biodiversity and vegetation dynamics in a forest ecosystem: case of djebel Boutaleb. Magister's thesis, Université Ferhat Abbas-Sétif, 91 p.

Troupin, G., (1988). Flora of Rwanda. Spermatophytes, volume 4.
Musée Royale de l'Afrique Centrale, Tervuren (Belgium), 651 pp.

Troupin, G., (1985). Flora of Rwanda. Spermatophytes, volume 3.
Musée Royale de l'Afrique Centrale, Tervuren (Belgium), 729 p.

Troupin, G., (1983). Flora of Rwanda. Spermatophytes, volume 2.
Royal Museum for Central Africa, Tervuren (Belgium), 603 p.

Troupin, G., (1978). Flora of Rwanda. Spermatophytes, volume 1. Musée Royale de l'Afrique Centrale, Tervuren (Belgium), 413 p.

IUCN, (2011). Burundi Parks and Reserves. Evaluation de l'efficacité of protected area management, 107 p.

Wibereho, W., (2010). Contribution à l'étude de la dynamique de la végétation du Parc National de la Kibira en zone de Bugarama; Mémoire d'Ingénieur Agronome; UB (FACAGRO), 105 p.

APPENDICES

Appendix 1: List and distribution of species collected site by site

N.B: For sites, 1 = Site Kaziramihunda 1, 2=Site Rwahirirwa, 3= Site Kaziramihunda 2, 4= Site Ndubura 1, 5= Site Ndubura 2, 6= Site Ndubura 3.

Families	Genres	Species	Common names	FB	EP	1	2	3	4	5	6
DICOTYLEDONES											
Acanthaceae	Mimulopsis	Mimulopsis solmsii Schweinf.	Umuna	P	Mo		+		+		
Acanthaceae	Brachystephanus	Brachystephanus africanus S. Moore		P	Mo						+
Acanthaceae	Brillantaisia	Brillantaisia cicatricosa Lindau	Ikinyakoko	P	SZ(O)		+	+	+		
Acanthaceae	Hypoestes	Hypoestes forskaolii(Vahl) Sol		P	Pal						+
Acanthaceae	Justicia	Justicia flava (Vahl) Vahl		Ch	Plur			+	+		
Acanthaceae	Isoglossa	Isoglossa gregorii (S.Moore) Lindau		Ch				+	+		
Acanthaceae	Anisosepalum	Anisosepalum humbertii (Mildbr.) E.Hossain	Akarishampongo	Ch	Mo		+				+
Alangiaceae	Alangium	Alangium chinense (L.f) Redher	Umukofo	P	Pal	+		+			+
Amaranthaceae	Sericostachys	Sericostachys scandens Gilg & Lopr.	Igitifu	P	Mo(EA)	2	+	2	+	2	
Annonaceae	Monanthotaxis	Monanthotaxis orophila (Boutique) Verdc.	Busase	P	Mo	+	+				+
Apocynaceae	Rauvolfia	Rauvolfia manni Staph	Ibamba	P	Mo	1	+	1		1	
Apocynaceae	Tabernaemontana	Tabernaemontana johnstonii (Staph) Pichon	Umudwedwe	P	Mo	2	+	3		2	+
Araliaceae	Schefflera	Schefflera goetzenii Harms	Ikizibakanwa	P	Mo	+	2	+			2
Araliaceae	Schefflera	Schefflera abyssinica (Hochst. Ex A. Rich.)	Umushorwe	P	Mo	1	+	+			
Araliaceae	Polyscia	Polyscia fulva (Hiern) Harms	Umwungo	P	Mo	2		2		2	+
Asclepiadaceae	Cynanchum	Cynanchum schistoglossum L.		Ch	SZ(OZ)	+		+		+	+
Asclepiadaceae	Rhynchostigma	Rhynchostigma racemosum Benth		P	Mo(EA)	+	+				
Asclepiadaceae	Tacazzea	Tacazzea apiculata Oliv.	Umunondo	P	Mo		+	+	+		
Aspidiaceae	Tectaria	Tectaria gemmifera (Fairy) Alston		G	Plur Afr			+			
Asteraceae	Gynura	Gynura scandens o. Hoffm	Ikidasha	P	Mo(EA)		+	+		1	+
Asteraceae	Mikaniopsis	Mikaniopsis usambarensis (Muschler) Milne-								+	
Asteraceae	Mikania	Mikania capensis DC.		P	Afr Mal	+				+	+
Asteraceae	Crassocephalum	Crassocephalum rubens (Juss.ex Jacq.) S. Moore	Igifurifuri	Th	Plur Afr		+	+			

114

Family	Genus	Species	Local name	Life form	Distribution						
Asteraceae	*Crassocephalum*	*Crassocephalum montuosum*(Juss.ex Jacq.) S. Moore	Igifurifuri	Ch	Afr Trop		+				
Asteraceae	*Sigesbeckia*	*Sigesbeckia orientalis* L.								+	
Asteraceae	*Adenostemma*	*Adenostemma viscosum* J.R. Forst. & G. Forst		Th	Subcosm			+			+
Asteraceae	*Senecio*	*Senecio mannii* (Hook.f.) C.Jeffrey	Umutagari	P	SZ-Mo				+		
Balsaminaceae	*Impatiens*	*Impatiens keilii* Gilg	Igisogoro								+
Balsaminaceae	Impatiens	*Impatiens burtonii* Hook.f.	Igisogoro	Th	SZ-Mo	1	2		1		
Balsaminaceae	Impatiens	*Impatiens stuhlmannii* Warb.	Igisogoro	Ch	SZ(Mo)	+		+	+		+
Balsaminaceae	Impatiens	*Impatiens purpureo-violacea* Gilg	Igisogoro	Ch	Mo(EA)	+	2	+			
Basellaceae	*Basella*	*Basella alba* L.	Inderama	P	Pan				+		1
Begoniaceae	*Begonia*	*Begonia meyeri-johannis* Engl.	Agafumbafumba	P	Mo	+	+	+	+	+	
Campanulaceae	*Canarina*	*Canarina eminii* Asch. & Schweinf.		E	Mo(EA)		+	+			
Caryophyllaceae	*Drymaria*	*Drymaria cordata* (L.) Willd.ex Roem & Schult.	Ururarwinzovu	Ch	Pan				+		
Celastraceae	*Gymnosporia*	*Gymnosporia acuminata* (L. f.) Szyszyl.		P	Plur	+	+	+		+	
Chrysobalanacea	*Magnistipula*	*Magnistipula butayei* De wild.	Umushwankima	P	SZ(OZ)	+		+			+
Chrysobalanacea	*Parinari*	*Parinari excelsa* Sabine	Umunazi	P	Plur Afr					+	
Clusiaceae	*Garcinia*	*Garcinia volkensii* Engl.	Umutundati							+	
Clusiaceae	*Symphonia*	*Symphonia globulifera* L.f.	Umushishi	P	Pan	1	1	+		+	+
Connaraceae	*Jaundea*	*Jaundea pinnata* (P.Beauv.) Schellenb.	Umuhasha	P	L.SZ-G	+	+	+		+	
Convolvulaceae	*Ipomoea*	*Ipomoea rubens* Choisy	Umurandaranda	Ch	Pal	+	+				
Crassulaceae	*Kalanchoe*	*Kalanchoe crenata* (Andrews) Haw.	Igitenetene	Ch	L.SZ-G	+		+			
Cucurbitaceae	*Momordica*	*Momordica foetida* Schumach.	Umwishwa	Ch	Plur Afr	+		+	+		+
Cucurbitaceae	*Coccinia*	*Coccinia mildbraedii* Harms	Icungucamuhare	P	SZ(O)		+	+		+	+
Cucurbitaceae	*Raphidiocystis*	*Raphidiocystis phyllocalyx* C.Jeffrey &		P	Mo	+	+		+		
Euphorbiaceae	*Macaranga*	*Macaranga kilimandscharica* pax	Umutwenzi	P	Mo(EA)	2	2	1		2	
Euphorbiaceae	*Alchornea*	*Alchornea hirtella* Benth.	Imvobo	P	L.SZ-G		1	1	+	3	
Euphorbiaceae	*Erythrococca*	*Erythrococca bongensis* Pax	Umutinti	P	SZ(Mo)			+	+	+	+
Euphorbiaceae	*Neoboutonia*	*Neoboutonia macrocalyx* Pax	Igihondogori	P	Mo			+	2		
Euphorbiaceae	*Bridelia*	*Bridelia brideliifolia* (Pax) Fedde	Umugimbu	P	SZ(Mo)			+			+
Euphorbiaceae	*Shirakiopsis*	*Shirakiopsis elliptica* (Hochst.) Esser	Umusasa	P	L.SZ-G		+				+
Euphorbiaceae	*Acalypha*	*Acalypha ornata* Hochst. ex A.Rich.		P	L.SZ-G				+		
Fabaceae	*Dalbergia*	*Dalbergia lactea* Vatke	Umuhasha	P	Afr Trop			+	+		
Fabaceae	*Desmodium*	*Desmodium repandum* (Vahl) D.C.		Ch	Pal		+	+		+	
Fabaceae	*Mimosa*	*Mimosa montana*	Uruzira					+			+

Family	Genus	Species	Vernacular name									
Fabaceae	*Acacia*	*Acacia montigena* Brenan	Umubambangwe					+				
Fabaceae	*Teramnus*	*Teramnus labialis* (L.f.) Sprengel		Ch	G			+				
Flacourtiaceae	*Lindackeria*	*Lindackeria kivuensis* Bamps	Umukundambazo	P	SZ(O)	+		+				
Geraniaceae	*Geranium*	*Geranium aculeolatum* Oliv.		Ch	SZ(EOZ)				+			
Hippocrateaceae	*Simirestis*	*Simirestis goetzei* (Loes.) N. Hallé ex R. Wilczek		P	SZ(OZ)			+		+		
Lamiaceae	*Plectranthus*	*Plectranthus melleri* Baker	Igifashi	Ch	Mo(EA)	+	+				+	2
Lamiaceae	*Achyrospermum*	*Achyrospermum micranthum* Perkins	Igifashi					+				2
Lamiaceae	*Isodon*	*Isodon ramosissimus* (Hook.f)Codd		Ch	Mo		+					2
Malvaceae	*Hibiscus*	*Hibiscus ludwigii* Eckl. & Zeyh.		P	SZ(EOZ)				+			
Meliaceae	*Lepidotrichilia*	*Lepidotrichilia volkensii* (Guerkke) Leroy	Umutana	P	SZ-Mo			+				
Melianthaceae	*Bersama*	*Bersama abyssinica* Fresen.	Umurerabana	P	Mo	+	+	+				
Menispermaceae	*Stephania*	*Stephania abyssinica* (Quart. Dill & A.Rich.)	Umuhanda	P	Mo	+	+	+	+			
Monimiaceae	*Xymalos*	*Xymalos monospora* (Harv.)	Umuhotora	P	Mo	2	1	1		1		
Moraceae	*Myrianthus*	*Myrianthus holstii* Engl.	Umwufe	P	Mo	2+		3		1		
Myrsinaceae	*Maesa*	*Maesa lanceolata* Forssk.	Umuhangahanga	P	Mo(EA)		+	+	+			+
Myrsinaceae	*Embelia*	*Embelia Schimperi* Vatke	Umukarakara	P	SZ(EOZ)	+	+	+				+
Myrsinaceae	*Embelia*	*Embelia libeniana* Taton		P	Mo(EA)	+	+	+			+	
Myrtaceae	*Syzygium*	*Syzygium parvifolium* (Engl.)Mildbr.	Umugoti	P	Mo(EA)	2	3	1		1		
Olacaceae	*Strombosia*	*Strombosia scheffleri* Engl.	Umushiga	P	Afr too	+					2	2
Oleaceae	*Jasminum*	*Jasminum pauciflorum* Benth.		P	G	+		+			+	
Oleaceae	*Jasminum*	*Jasminum schimperi* Vatke									+	+
Oleaceae	*Jasminum*	*Jasminum abyssinicum* Hochst.ex DC.		P	SZ(EOZ)						+	
Passifloraceae	*Adenia*	*Adenia bequaertii* Robyns	Umubururangwe	P	Mo				+			
Phytolaccaceae	*Phytolacca*	*Phytolacca dodecandra* L'Hér.	Umwokora	P	Afr Mal				+			
Piperaceae	*Piper*	*Piper capense* L.f	Umukonjoro	P	Plur Afr	+		+	+			
Piperaceae	*Peperomia*	*Peperomia tetraphylla* (G.Forst.) Hook & Arn.		Ch	Pan	+	+	+		+		
Piperaceae	*Peperomia*	*Peperomia fernando-poiana* C.DC.		Th	Mo	+		+				+
Piperaceae	*Peperomia*	*peperomia blanda* (Jacq.) Kunth				+		+				
Polygonaceae	*Rumex*	*Rumex abyssinicus* Jacq.	Igifumbafumba	G	Afr Mal				+			
Ranunculaceae	*Clematis*	*Clematis simensis* Fresen.		P	SZ				+			1
Ranunculaceae	*Thalictrum*	*Thalictrum rhynchocarpum* Dillon ex A.Rich.		H	Mo				+			
Rhamnaceae	*Gouania*	*Gouania longispicata* Engl.	Umubimbafuro	P	L.SZ-G			+				+
Rosaceae	*Prunus*	*Prunus africana* (Hook.f.) Kalkman	Umuremera	P	Plur				2			+

Family	Genus	Species	Vernacular name	Habit	Distribution						
Rosaceae	Rubus	Rubus pinnatus Willd.	Umukere	P	Mo	+	+		+		
Rosaceae	Rubus	Rubus steudneri Schweinf.	Umukere	P	Mo(EA)				+		
Rubiaceae	Galiniera	Galiniera saxifraga(Hochst.) Bridson	Ikiryoheramuhoro	P	Mo	1	1	+	+	+	
Rubiaceae	Psychotria	Psychotria palustris E.M.A.Petit	Ikiryohera	P	End	2	2	1	+	+	
Rubiaceae	Psychotria	Psychotria bugoyensis K.Krause	Ikiryohera	P	Mo(EA)						+
Rubiaceae	Psychotria	Psychotria sp. 1	Ikiryohera						+		
Rubiaceae	Chassalia	Chassalia subochreata (De Wild.) Robyns	Umukotambugita	P	Mo(EA)	1	+	+		2	
Rubiaceae	Virectaria	Virectaria major (K.Schum.) Verdc.	Umukizikizi	Th	Mo		+				
Rubiaceae	Oxyanthus	Oxyanthus speciosus DC.	Umutore	P	Afr too	+		+			+
Rubiaceae	Keetia	Keetia gueinzii(Sond.) Bridson				+	1	1			
Rubiaceae	Rytigynia	Rytigynia bridsoniiVerdc.		P	Mo(EA)		+	+			1
Rubiaceae	Rytigynia	Rytigynia kiwuensis (K.Krause) Robyns	Umukondokondo	P	Mo(EA)	+	+	+			
Rubiaceae	Sabicea	Sabicea venosa Benth.					+	+		+	+
Rubiaceae	Vangueria	Vangueria apiculata K.Schum.	Umutobo			+		+			1
Rubiaceae	Pauridiantha	Pauridiantha paucinervis (Hiern) Bremek.	Umusivya	P	Mo	+			1		
Rubiaceae	Sericanthe	Sericanthe burundensis Robbr.				+				+	
Rubiaceae	Rubiaceae Indet	Rubiaceae 1								+	+
Rubiaceae	Rubiaceae Indet	Rubiaceae 2								+	+
Rubiaceae	Rubiaceae Indet	Rubiaceae 3								+	+
Rutaceae	Vepris	Vepris renieri (G.C.C. Gilbert) Mziray	Umuzonyarubabi	P	End	+				+	2
Sapindaceae	Allophylus	Allophylus chaunostachys Gilg	Umuvumereza	P	Mo	+		+		+	
Sapindaceae	Allophylus	Allophylus ferrugineus Taub. var. ferrugineus	Umuvumereza	P	SZ(OZ)		+				2
Sapindaceae	Sapindaceae Indet	Sapindaceae 1							+		
Sapotaceae	Chrysophyllum	Chrysophyllum gorungosanum Engl.	Umuko	P	Mo			1		3	+
Sapotaceae	Afrosersalisia	Afrosersalisia rwandensis (Troupin) Liben	Umuko	P	Mo(EA)	+					
Sapotaceae	Synsepalum	Synsepalum seretii (De Wild.) T. D. Penn.								+	
Solanaceae	Solanum	Solanum nigrum L.	Insogo	Th	Cos				+		
Solanaceae	Discopodium	Discopodium penninervium Hochst.	Umurenga	P	Mo				+		+
Sterculiaceae	Dombeya	Dombeya goetzenii K.Schum.	Umukore	P	SZ(O)				1		+
Theaceae	Balthasaria	Balthasaria schliebenii (Melchior) Verdcourt	Umusasa			+	+	+			+
Tiliaceae	Triumfetta	Triumfetta cordifolia A.Rich.	Umusarenda	P	Afr too		+		+		
Urticaceae	Laportea	Laportea alatipes Hook.f.	Igisuru	Th	Mo			+			
Urticaceae	Urera	Urera hypselodendron (Hochst.ex A.Rich.) Wedd	Umumbiri	P	Afr Mal	1	+	+	+	+	

117

Family	Genus	Species	Local name								
Urticaceae	*Pilea*	*Pilea bambuseti* Engl.	Umunzenge	Th	Mo			+		1	
Urticaceae	*Elatostema*	*Elatostema monticola* Hook.f.						+			
Urticaceae	*Laportea*	*Laportea ovalifolia* (Schum.) Chew						+	+		
Urticaceae	*Urtica*	*Urtica massaica* mildbr.		H	Mo			+			+
Verbenaceae	*Clerodendrum*	*Clerodendrum johnstonii* Oliv	Umunyankuru	P	SZ-Mo	+					
Verbenaceae	*Clerodendrum*	*Clerodendrum buchlolzii* GÜrke	Umugutabateme	P	Mo						+
Verbenaceae	*Clerodendrum*	*Clerodendrum* Sp. 1							+	+	
Verbenaceae	*Clerodendrum*	*Clerodendrum* Sp. 2				+		+			
Verbenaceae	*Clerodendrum*	*Clerodendrum* Sp. 3								+	
Verbenaceae	*Clerodendrum*	*Clerodendrum* Sp. 4									+
Vitaceae	*Cyphostemma*	*Cyphostemma mildbraedii* Gilg & M.Brandt	Agasharita	G	SZ(Z)	+		+	+		
MONOCOTYLEDONES											
Anthericaceae	*Chlorophytum*	*Chlorophytum sparsiflorum* Baker				+					+
Araceae	*Culcasia*	*Culcasia falcifolia* Engl.	Intaruka	P	Plur Afr	+		+	+	+	
Araceae	*Arisaema*	*Arisaema mildbraedii* Engl.		G	Mo(EA)	+	+	+			+
Commelinaceae	*Commelina*	*Commelina* sp. 1	Uruteza					+			+
Cyperaceae	*Cyperus*	*Cyperus pseudoleptocladus* kÜk.	Ikigaga	Ch	SZ(O)	+	+	+			+
Cyperaceae	*Cyperus*	*Cyperu* sp. 1	Ikigaga			+	+	+	+		
Dracaenaceae	*Dracaena*	*Dracaena afromontana* Mildbr.	Igishwankenke	P	Mo	1	1	1	+	2	
Musaceae	*Enste*	*Ensete ventricosum* (Welw.) Cheesman	Ikigomogomo	P	Mo			+		1	
Orchidaceae	*Diaphanathe*	*Diaphanathe bilobata* (Summerh.) Rasmussen.				+	+	+			
Orchidaceae	Habenaria	*Habenaria brachylobos* (Summerh.) Summerh.						+			
Orchidaceae	*Chamaeangis*	*Chamaeangis vesicata* (Lindl.) Schltr.						+			1
Orchidaceae	*Aerangis*	*Aerangis ugandensis* Summerh.		E	SZ(O)			+			
Orchidaceae	*Polystachya*	*Polystachya virginea* Summerh.		Ch	Mo(EA)			+			
Orchidaceae	*Polystachya*	*Polystachya gracilenta* Kraenzl.					+				
Orchidaceae	*Polystachya*	*Polystachya cultriformis* (Trouars) Lindl. ex		H	Afr Mal			+			
Poaceae	*Setaria*	*Setaria megaphylla* (Steud.)T.Durand	Igikaranka	H	L.SZ-G	+	+		+		
Poaceae	*Isachne*	*Isachne mauritiana* Kunth	Inyegeshi	Ch	Afr Mal	+	+	+			
Smilacaceae	*Smilax*	*Smilax anceps* Willd.	Umusuri	P	Plur Afr	+	+	+			
PTERIDOPHYTES											
Aspleniaceae	*Asplenium*	*Asplenium dregeanum* Kunze	Agashurushuru	G	Plur Afr			+			
Aspleniaceae	*Asplenium*	*Asplenium mildbraedii* Hieron.	Agashurushuru							1	1

Family	Genus	Species	Location								
Aspleniaceae	*Asplenium*	*Asplenium elliotii* C.H. Wright	Agashurushuru	G	Mo	+	+	1		+	
Aspleniaceae	*Asplenium*	*Asplenium sandersonii* Hook.	Agashurushuru	H	Afr Mal		1	+			
Aspleniaceae	*Asplenium*	*Asplenium aethiopicum* (Burm.f.) Bech.	Agashurushuru	G	Pal			1			+
Aspleniaceae	*Asplenium*	*Asplenium glamosum* Willd	Agashurushuru					+			+
Aspleniaceae	*Asplenium*	*Asplenium mannii* Hook.	Agashurushuru	G	Afr Trop					1	2
Aspleniaceae	*Asplenium*	*Asplenium friesiorum* C.Chr.	Agashurushuru	G	Plur Afr		+				1
Cyatheaceae	*Cyathea*	*Cyathea dregei* Kunze	Mugogutarengwa	P				+			
Cyatheaceae	*Cyathea*	*Cyathea manniana* Hook.	Mugogutarengwa			+		1	+	+	+
Dennstaedtiaceae	*Pteridium*	*Pteridium aquilinum* (L.) Kuhn	Igishurushuru	G	Cos				+		
Dryopteridaceae	*Dryopteris*	*Dryopteris kilemensis* (Kuhn) Kuntze	Iraba				1	+		+	
Dryopteridaceae	*Dryopteris*	*Dryopteris pentheri* (Krasser) C.Chr.	Iraba	G	Plur Afr	+	+	+			
Dryopteridaceae	*Dryopteris*	*Dryopteris* sp. 1						+	+		
Oleandraceae	*Arthropteris*	*Arthropteris monocarpa* Auct.non (Cordem.)					+	+		+	1
Polypodiaceae	*Loxogram*	*Loxogramma abyssinica* (Baker) Price		G	Afr Mal	+	+	+			+
Pteridaceae	*Pteris*	*Pteris kivuensis* C.Chr.				+		+			+
Vittariaceae	*Vittaria*	*Vittaria reekmansii* Pic. Serm.				+					+
BRYOPHYTES											
Hymenophyllace	*Crepidomanes*	*Crepidomanes inopinatum* (Pic. Serm.) J.P. Roux				+		+			+

Appendix 2: Dominant families by site

Species by family and site	Kaziramihunda 1	Rwahirirwa	Kaziramihunda 2	Ndubura 1	Ndubura 2	Ndubura 3	Total
Families							
Broadleaf	**21**	**21**	**39**	**22**	**18**	**23**	**60**
Acanthaceae	0	3	3	4	0	4	7
Asteraceae	1	2	3	3	4	3	7
Balsaminaceae	3	2	2	2	0	1	4
Euphorbiaceae	1	3	5	4	3	3	7
Fabaceae	0	1	5	1	1	1	5
Piperaceae	4	1	4	1	1	1	4
Rubiaceae	9	8	10	3	6	5	14
Urticaceae	1	1	6	3	1	2	6
Verbenaceae	2	0	1	1	2	3	6
Monocotyledons	**1**	**2**	**6**	**0**	**0**	**0**	**7**
Orchidaceae	0	2	6	0	0	0	7
Pteridophytes	**2**	**3**	**5**	**1**	**2**	**3**	**8**
Aspleniaceae	2	3	5	1	2	3	8
Total	**24**	**26**	**50**	**23**	**20**	**26**	**75**

120

Appendix 3: Tables showing the distribution of stems by circumference class for the different sites Table 1: Kaziramihunda 1 site

Circumference classes	Macaranga neomildbraediana	Tabernaemontana johnstonii	Xymalos monospora	Galiniera coffeoides	Psychotria palustris	Magnistipula butayei	Rytigynia Kivuensis	Dracaena afromontana	Bersama abyssinica	Polyscia fulva	Lindackeria kivuensis	Chassalia subochreata	Indét	Shirakiopsis elliptica	Myrianthus holstii	Syzygium parvifolium	Cyathea sp	Total	percentage
[15-25[	12	22	13	20	11	3	8	1	1	0	2	1	0	0	4	3	2	103	37,1
[25-35[	27	18	8	3	2	1	1	1	0	0	0	0	1	0	2	1	3	68	24,5
[35-45[	11	7	10	2	2	2	0	1	1	0	0	0	0	1	1	0	0	38	13,7
[45-55[	10	7	0	0	0	1	1	1	0	0	0	0	0	0	1	1	0	22	7,9
[55-65[	7	4	3	0	1	0	0	2	0	0	0	0	0	0	2	0	0	19	6,8
[65-75[	4	2	1	0	0	0	0	0	1	0	0	0	0	0	1	0	0	9	3,2
[75-85[	1	0	1	1	0	1	0	0	0	0	0	0	0	0	1	1	0	6	2,2
[85-95[	1	0	3	0	0	0	0	0	0	0	0	0	0	0	0	1	0	5	1,8
[95-105[	0	0	0	0	0	0	0	0	0	0	0	0	0	0	0	1	0	1	0,4
[105-115[	1	0	0	0	0	0	0	0	0	0	0	0	0	0	0	0	0	1	0,4
[115-125[	0	0	1	0	0	0	0	0	0	0	0	0	0	0	0	0	0	1	0,4
[125-135[	0	0	0	0	0	1	0	0	0	0	0	0	0	0	0	1	0	2	0,7
[145-155[	0	0	0	0	0	0	0	0	0	0	0	0	0	1	0	0	0	1	0,4
[175-185[	0	0	0	0	0	0	0	0	0	1	0	0	0	0	0	0	0	1	0,4
[215-225[	0	0	0	0	0	0	0	0	0	1	0	0	0	0	0	0	0	1	0,4
Total	74	60	40	26	16	9	10	6	3	2	2	1	1	2	12	9	5	278	100
Percentage	26,6	21,6	14,4	9,4	5,8	3,2	3,6	2,2	1,1	0,7	0,7	0,4	0,4	0,7	4,3	3,2	1,8	100	

Table 2: Rwahirirwa site

Circumference classes	*Galiniera coffeoides*	*Macaranga kilimandscharica*	*Xymalos monospora*	*Syzygium parvifolium*	*Psychotria palustris*	*Dracaena afromontana*	Indét	*Tabernaemontana johnstonii*	*Symphonia globulifera*	*Maesa lanceolata*	*Myrianthus holstii*	*Bersama abyssinica*	*Cyathea maniana*	Total	%
[15-25[	13	3	3	1	6	0	0	0	0	0	0	0	1	27	18
[25-35[	11	7	2	0	1	0	0	0	1	0	1	0	17	40	27
[35-45[	15	5	4	1	0	0	0	0	0	0	0	0	16	41	28
[45-55[	1	5	5	1	0	0	0	0	1	1	0	0	0	14	9,5
[55-65[	0	7	1	0	0	0	0	0	0	0	0	1	0	9	6,1
[65-75[	0	3	2	0	0	1	0	1	0	0	0	0	0	7	4,8
[75-85[	0	1	0	0	0	0	0	0	0	0	0	0	0	1	0,7
[85-95[	0	0	0	1	0	0	0	0	0	0	0	0	0	1	0,7
[95-105[	0	1	0	0	0	0	0	0	0	0	0	0	0	1	0,7
[105-115[	0	0	0	1	0	0	0	0	0	0	0	0	0	1	0,7
[135-145[	0	0	0	1	0	0	0	0	0	0	0	0	0	1	0,7
[145-155[	0	0	0	1	0	0	0	0	0	0	0	0	0	1	0,7
[165-175[	0	0	0	0	0	0	1	0	0	0	0	0	0	1	0,7
[205-215[	0	0	0	1	0	0	0	0	0	0	0	0	0	1	0,7
[265-275[	0	0	0	1	0	0	0	0	0	0	0	0	0	1	0,7
Total	40	32	17	9	7	1	1	1	2	1	1	1	34	147	100
%	27,21	21,77	12	6,1	4,762	0,7	0,7	0,7	1,4	0,7	0,68	0,68	23	100	

Table 3: Kaziramihunda 2 site

Circumference classes	*Myrianthus holstii*	*Tabernaemontana johnstonii*	*Xymalos monospora*	*Dracaena afromontana*	*Magnistipula butayei*	*Acacia montigena*	*Chassaria subochreata*	*Rytigynia kivuensis*	*Prinus africana*	Indet 1	*Polyscia fulva*	*Schefflera abyssinica*	Indet 2	*syzygium parvifolium*	*Pauridiantha paucinervis*	*Alchornea hirtella*	*Cyathea maniana*	Total	%
[15-25[	21	5	6	9	2	1	6	8	0	1	0	0	0	0	0	1	0	60	35,3
[25-35[	12	1	4	3	0	1	0	3	0	0	0	1	1	0	0	0	1	27	15,9
[35-45[	7	4	4	2	0	0	0	0	0	0	0	0	0	0	1	0	0	18	10,6
[45-55[	7	2	7	1	0	0	0	0	0	0	0	1	0	0	0	0	0	18	10,6
[55-65[	3	4	2	0	0	0	0	0	0	0	0	0	0	0	0	0	0	9	5,3
[65-75[	0	3	0	1	0	0	0	0	0	0	0	0	0	0	1	0	0	4	2,4
[75-85[	2	1	2	0	0	0	0	0	0	0	0	0	0	0	1	0	0	6	3,5
[85-95[	3	1	0	0	0	0	0	0	0	0	0	0	0	1	0	0	0	5	2,9
[95-105[	2	1	0	1	0	0	0	0	0	0	0	0	0	0	0	0	0	4	2,4
[105-115[	1	2	1	1	0	0	0	0	0	0	0	0	0	0	0	0	0	5	2,9
[115-125[	2	1	0	0	0	0	0	0	0	0	0	0	0	0	1	0	0	3	1,8
[145-155[	0	1	0	0	0	0	0	0	0	0	0	0	0	0	0	0	0	1	0,6
[155-165[	0	0	0	0	0	0	0	0	0	0	0	0	0	0	1	0	0	1	0,6
[175-185[	0	1	0	0	0	0	0	0	1	0	0	0	0	0	0	0	0	2	1,2
[195-205[	0	1	0	0	0	0	0	0	0	0	0	0	0	0	0	0	0	1	0,6
[215-225[	0	0	0	0	0	0	0	0	0	0	0	0	0	0	1	0	0	1	0,6
[235-245[	0	0	0	0	0	0	0	0	0	0	1	0	0	0	0	0	0	1	0,6
[285-295[	0	0	0	0	0	0	0	0	0	0	0	0	0	0	1	0	0	1	0,6

																			Total	%
[305-315[	0	0	0	0	0	0	0	0	0	0	1	0	0	0	0	0	0	1	0,6
[375-385[	0	0	0	0	0	0	0	0	1	0	0	0	0	0	0	0	0	1	0,6
[415-425]	0	0	0	0	0	0	0	0	1	0	0	0	0	0	0	0	0	1	0,6
Total	60	28	26	18	2	2	6	11	3	1	2	2	1	1	5	1	1	170	100
%	35,3	16,5	15,3	10,6	1,2	1,2	3,5	6,5	1,8	0,6	1,2	1,2	0,6	0,6	2,9	0,6	0,6	100	

Table 4: Ndubura 1 site

	Neoboutonia macrocalyx	*Dombeya goetznii*	*Discopodium penninervium*	Total	%
[15-25[	4	0	11	15	48,39
[25-35[	4	1	0	5	16,13
[35-45[	1	0	0	1	3,23
[45-55[	1	1	0	2	6,45
[55-65[	1	0	0	1	3,23
[65-75[	4	1	0	5	16,13
[85-95[	2	0	0	2	6,45
Total	17	3	11	31	
%	54,84	9,68	35,48		100

Table 5: Ndubura 2 site

Circumference classes	Macaranga kilimandscharica	Myrianthus holstii	Tabernaemontana johnstonii	Pauridiantha paucinervis	Chassaria subochreata	Alchornea hirtella	Chrysophyllum gorungosanum	Strombosia Scheffleri	Xymalos monospora	Synsepalum seretii	Magnistipula butayei	Rauvolfia mamii	Dracaena afromontana	Psychotria palustris	Sericanthe burundensis	Symphonia globulifera	Indet	Polyscia fulva	Total	%
[15-25[	9	1	2	0	2	28	0	0	1	0	1	1	3	2	3	2	0	0	55	50,93
[25-35[	5	0	1	0	0	3	0	0	2	0	0	0	0	0	0	0	0	0	11	10,19
[35-45[	3	0	0	1	0	0	0	0	0	0	0	0	1	0	0	1	0	0	6	5,56
[45-55[	0	1	0	1	0	0	0	0	0	0	0	0	0	0	0	0	0	0	2	1,85
[55-65[	1	3	1	0	0	0	1	1	0	0	0	0	0	0	0	0	0	0	7	6,48
[65-75[	0	0	0	0	0	0	0	1	0	0	0	0	0	0	0	0	0	0	1	0,93
[75-85[	2	0	2	0	0	0	0	0	1	1	0	0	1	0	0	0	0	0	7	6,48
[85-95[	0	0	0	0	0	0	0	1	0	0	0	0	0	0	0	0	0	0	1	0,93
[95-105[	1	1	0	0	0	0	0	0	1	0	0	0	0	0	0	0	0	0	3	2,78
[105-115[	0	0	0	0	0	0	0	1	0	0	0	0	0	0	0	0	0	0	1	0,93
[115-125[	0	0	1	0	0	0	0	1	0	0	0	0	0	0	0	0	1	0	3	2,78
[125-135[	0	0	0	0	0	0	0	0	0	1	0	0	0	0	0	0	0	0	1	0,93
[135-145[	0	1	0	0	0	0	1	0	0	0	0	0	0	0	0	0	0	0	2	1,85
[155-165[	1	0	0	0	0	0	0	0	0	0	0	0	0	0	0	0	0	0	1	0,93
[175-185[	0	0	0	0	0	0	1	0	0	0	0	0	0	0	0	0	0	0	1	0,93
[195-205[	0	0	0	0	0	0	1	0	0	1	0	0	0	0	0	0	0	0	2	1,85
[215-225[	0	1	0	0	0	0	0	0	0	0	0	0	0	0	0	0	0	1	2	1,85

																			Total	
[255-265[	0	0	0	0	0	0	1	0	0	0	0	0	0	0	0	0	0	0	1	0,93
[265-275[	0	0	0	0	0	0	1	0	0	0	0	0	0	0	0	0	0	0	1	0,93
Total	22	8	7	2	2	31	6	5	5	3	1	1	5	2	3	3	1	1	108	
%	20,37	7,41	6,48	1,85	1,85	28,70	5,56	4,63	4,63	2,78	0,93	0,93	4,63	1,85	2,78	2,78	0,93	0,93		100

Table 6: Ndubura 3 site

Circumference classes	*Macaranga kilimandscharica*	*Chrysophyllum gorungosanum*	*Myrianthus holstii*	*Strombosia scheffleri*	*Erythrococca bongensis*	*Psychotria palustris*	*Rytigynia bridsonii*	*Garcinia volkensii*	*Tabernaemontana johnstonii*	*Dracaena afromontana*	*Symphonia globulifera*	*Maesa lanceolata*	*Xymalos monospora*	*Neoboutonia macrocalyx*	*Magnistipula butayei*	*Syzygium parvifolium*	Total	%
[15-25[	0	0	3	6	0	2	2	13	2	3	0	0	0	0	1	0	32	32,99
[25-35[	1	1	4	4	1	0	0	2	2	0	1	0	0	0	0	0	16	16,49
[35-45[	0	2	2	1	0	0	0	0	2	0	0	1	0	0	0	0	8	8,25
[45-55[	1	0	1	0	0	0	0	0	0	0	1	0	0	1	0	0	4	4,12
[55-65[	0	1	2	0	0	0	0	0	0	1	0	0	0	0	0	0	4	4,12
[65-75[	1	0	1	0	0	0	0	0	1	0	0	0	0	0	0	0	3	3,09
[75-85[	1	1	0	2	0	0	0	0	1	0	0	0	1	0	0	0	6	6,19
[85-95[	0	2	0	1	0	0	0	0	1	0	0	0	0	0	0	0	4	4,12
[95-105[	0	1	0	0	0	0	0	0	2	0	0	0	0	0	0	0	3	3,09
[105-115[	0	0	0	1	0	0	0	0	1	0	0	0	0	0	0	0	2	2,06
[115-125[	1	0	1	0	0	0	0	0	1	0	0	0	0	0	0	0	3	3,09
[125-135[	0	1	0	0	0	0	0	0	1	0	0	0	0	0	0	0	2	2,06
[135-145[	0	1	0	0	0	0	0	0	0	0	0	0	0	0	0	1	2	2,06

[155-165[	1	1	0	0	0	0	0	0	0	0	0	0	0	0	0	0	2	2,06
[165-175[	0	1	0	0	0	1	0	0	0	0	0	0	0	0	0	0	2	2,06
[185-195[	0	1	0	1	0	0	0	0	0	0	0	0	0	0	0	0	2	2,06
[255-265[	0	0	1	0	0	0	0	0	0	0	0	0	0	0	0	0	1	1,03
[305-315[	0	0	0	0	0	0	0	0	0	0	1	0	0	0	0	0	1	1,03
Total	6	13	15	16	1	3	2	15	14	4	3	1	1	1	1	1	97	
%	6,19	13	15	16	1	3,1	2,1	15	14	4,1	3,1	1,03	1,03	1	1,03	1		100

Appendix 4: Recording form for observations of plant dynamics habitats of Burundi's protected areas (Habiyaremye, 2012)

Contributeur :

SITE :															
Référence des observations						Activités humaines		Faune			Observations Habitats				
N	Date	Position				Type	Surf	Signe	Espèces	Nbr	Type	Strates	Ht	Rc	Espèces	Ph
		GPS			Topo										
		Alt	Lat	Lon											
												A-TGA			- - - - -	
												A-GA			- - - - -	
												A-AM			- - - - -	
												aB			- - - - -	
												SsAH			- - - - -	

Légende (explication des abréviations)

Référence des observations	Activités humaines	Faune	Observations habitats	
Hre : heure Lat : latitude Lon : longitude N : numéro Topo : toponyme	Surf : surface (m²)	Nbr : Nombre	**Strates**	**Hauteur (m)**
			•A-TGA: arborescente avec de très grands arbres	30 < 50
			•A-GA: arborescente avec de grands arbres	20 < 30
			•A-AM: arborescente constituée d'arbres petits à moyens	7 < 20
			•aB: arbustive	2 < 7
			•SsAH: sous-arbustive et/ou herbacée	< 2
			Ht : hauteur (m)	
			Rc : recouvrement (%)	
			Ph : photo	

Appendix 5: Tree dendrometric measurements

Species by site	Individuals	Circumference (Cm)	Height (m)	Species by site	Individuals	Circumference (Cm)	Height (m)
Kaziramihunda 1				**Kaziramihunda 2**			
Macaranga kilimandscharica	1	113	14	*Myrianthus holstii*	1	88	10
	2	45	12		2	51	8
	3	55	16		3	18	4
	4	18	11		4	23	3.5
	5	56	14		5	16	3
	6	27	7		6	24	4
	7	74	12		7	29	5
	8	39	12		8	20	5
	9	58	15		9	51	10
	10	18	4		10	32	9
	11	47	14		11	26	5
	12	56	16		12	19	7
	13	46	12		13	15	2.5
	14	50	12		14	15	3
	15	30	12		15	52	15
	16	89	20		16	24	4
	17	37	12		17	37	10
	18	37	14		18	37	8
	19	58	14		19	118	7
	20	35	11		20	120	7
	21	25	11		21	38	7
	22	32	11		22	17	4
	23	28	8		23	24	5
	24	33	10		24	33	8
	25	26	10		25	28	7
	26	29	11		26	16	3.5
	27	23	10		27	27	6
	28	19	14		28	21	4
	29	40	13		29	32	6
	30	42	14		30	36	6
	31	45	15		31	19	4
	32	48	15		32	27	4
	33	27	14		33	25	5
	34	52	15		34	99	18
	35	58	19		35	44	10
	36	45	13		36	53	12
	37	28	9		37	104	17
	38	31	15		38	49	1.5
	39	48	15		39	32	6
	40	28	10		40	22	5
	41	44	13		41	25	5
	42	73	20		42	16	2
	43	26	11		43	16	2.5
	44	84	22		44	15	2

Species	No.			Species	No.		
	45	34	13		45	16	3
	46	26	13		46	89	13
	47	63	18		47	78	20
	48	65	19		48	63	12
	49	27	9		49	29	7
	50	30	8		50	41	8
	51	34	12		51	41	8
	52	25	11		52	50	7
	53	16	8		53	80	10

Species	No.			Species	No.		
	54	23	10		54	47	8
	55	42	14		55	61	16
	56	47	14		56	89	17
	57	24	10		57	22	4
	58	34	11		58	64	14
	59	21	13		59	105	20
	60	24	11		60	22	4
	61	22	9	*Tabernaemontana johnstonii*	1	28	7
	62	38	14		2	55	9
	63	30	12		3	38	7
	64	27	12		4	57	9
	65	30	11		5	38	9
	66	15	7		6	71	8
	67	29	11		7	124	20
	68	42	15		8	71	12
	69	28	13		9	20	12
	70	30	13		10	41	9
	71	20	9		11	16	4.5
	72	27	8		12	46	6
	73	37	12		13	21	3
	74	25	12		14	22	3
Tabernaemontana johnstonii	1	29	8		15	21	3
	2	42	10		16	179	25
	3	22	7		17	47	15
	4	20	4		18	103	15
	5	74	12		19	44	11
	6	30	6		20	110	20
	7	15	4		21	108	20
	8	51	11		22	94	18
	9	46	11		23	74	20
	10	31	10		24	63	12
	11	28	9		25	83	12
	12	51	11		26	153	17

Species	No.			Species	No.		
	13	42	28		27	195	18
	14	36	9		28	59	14
	15	25	6	*Xymalos monospora*	1	26	4
	16	47	12.5		2	46	11
	17	62	15		3	21	3
	18	29	6		4	50	5
	19	16	4		5	38	7
	20	23	7		6	23	3
	21	21	6		7	25	7
	22	17	6		8	16	2
	23	18	5		9	80	17

Species	No.			Species	No.		
	24	31	10		10	25	5
	25	52	9		11	60	13
	26	31	4		12	59	12
	27	31	4		13	21	12
	28	21	5		14	47	12
	29	33	9		15	23	6
	30	68	10		16	31	2
	31	59	9		17	20	4
	32	23	2.5		18	43	7
	33	63	12		19	112	10
	34	61	9		20	54	12
	35	46	9		21	80	14
	36	18	5		22	46	7
	37	18	5		23	49	12
	38	28	8		24	41	6
	39	24	5		25	44	8
	40	27	6		26	47	8
	41	32	8	*Dracaena afromontana*	1	21	5
	42	51	10		2	16	3
	43	29	6		3	18	4
	44	39	5		4	103	12
	45	32	7		5	25	5

Species	No.			Species	No.		
	46	24	7		6	35	8
	47	16	4		7	27	4
	48	34	9		8	14	3
	49	23	7		9	16	3
	50	24	4		10	19	3
	51	40	8		11	42	6
	52	38	8		12	113	3.5
	53	34	7		13	15	3
	54	24	7		14	47	8
	55	38	9		15	71	9
	56	21	5		16	19	2.5
	57	27	8		17	29	3
	58	15	5		18	23	3
	59	24	6	*Magnistipula butayei*	1	19	7
	60	16	3		2	18	7
Xymalos monospora	1	17	3.5	*Acacia montigena*	1	31	20
	2	20	5		2	22	20
	3	79	15	*Chassalia subochreata*	1	16	3
	4	64	12		2	18	3
	5	85	18		3	16	2
	6	42	6		4	20	4
	7	93	18		5	18	3
	8	26	6		6	16	2
	9	62	9	*Rytigynia kiwuensis*	1	17	3
	10	37	7		2	19	7
	11	36	7		3	16	4
	12	31	6		4	22	4
	13	17	6		5	22	4
	14	15	7		6	16	3
	15	115	12		7	28	3
	16	42	9		8	15	3
	17	28	6		9	25	5
	18	15	6		10	26	5

	19	16	6		11	21	3
	20	31	8	*Prunus africana*	1	381	40
	21	90	8		2	182	29
	22	31	8		3	420	30
	23	42	9	Indét	1	16	7
	24	25	8	*Polyscias fulva*	1	310	25
	25	35	5		2	241	30
	26	20	14	*Schefflera abyssinica*	1	46	18
	27	15	9		2	32	18
	28	18	6	Indét	1	29	19
	29	21	6	*Syzygium parvifolium*	1	158	20
	30	19	8		2	220	30
	31	27	5		3	85	22
	32	43	8		4	286	40
	33	61	7	*Pauridiantha paucinervis*	1	21	3
	34	17	6		2	22	2
	35	68	9		3	21	7
	36	42	9		4	30	10
	37	35	7		5	22	12
	38	29	6	*Alchornea hirtella*	1	22	4
	39	17	12	*Cyathea maniana*	1	34	4
	40	42	12	**Ndubura 1**			
Galiniera saxifraga	1	28	7	*Neoboutonia macrocalyx*	1	67	12
	2	83	9		2	73	12
	3	19	7		3	62	19
	4	21	6		4	46	9
	5	19	6		5	23	5
	6	24	5		6	31	5
	7	17	7		7	26	4
	8	19	5		8	36	6
	9	17	5		9	20	3.5
	10	24	4		10	15	1.5
	11	15	6		11	18	4

	12	18	4		12	67	14
	13	17	6		13	25	4
	14	32	7		14	32	5
	15	18	7		15	85	13
	16	21	5		16	72	9
	17	23	4		17	93	10
	18	18	3.5	*Dombeya goetznii*	1	30	6
	19	18	6		2	49	8
	20	43	7		3	71	3
	21	30	8	*Discopodium pennninervium*	1	15	2
	22	42	8		2	14	3
	23	16	5		3	18	3
	24	19	5		4	18	3
	25	18	7		5	15	2.5
	26	19	6		6	19	4
Psychotria palustris	1	17	6		7	16	3
	2	16	4		8	16	3
	3	56	10		9	17	4
	4	25	9		10	20	4
	5	15	4		11	19	4
	6	37	6	**Ndubura 2**			
	7	15	7	*Macaranga kilimandscharica*	1	84	20

	8	16	9		2	103	25
	9	16	6		3	159	27
	10	17	7		4	25	6
	11	23	7		5	41	12
	12	18	8		6	25	8
	13	22	5		7	22	7
	14	20	5		8	16	5
	15	43	10		9	28	10
	16	25	6		10	31	7
Magnistipula butayei	1	77	15		11	27	10
	2	15	9		12	17	8

	3	21	8		13	23	10
	4	18	7		15	64	15
	5	41	13		16	22	10
	6	42	15		17	43	15
	7	46	12		18	15	3
	8	32	15		19	15	3
	9	132	22		20	78	20
Rytigynia kiwuensis	1	21	6		21	22	9
	2	21	7		22	20	3
	3	17	6	*Myrianthus holstii*	1	101	18
	4	15	5		2	141	27
	5	17	6		3	56	12
	6	18	7		4	61	16
	7	21	8		5	19	4
	8	29	4		6	50	5
	9	20	7		7	63	7
	10	52	11		8	215	25
Dracaena afromontana	1	51	11	*Tabernaemontana johnstonii*	1	84	15
	2	20	7		2	58	14
	3	40	4		3	22	5
	4	31	3		4	19	27
	5	57	9		5	82	15
	6	61	9		6	32	8
Bersama abyssinica	1	70	14		7	123	16
	2	41	4	*Pauridiantha paucinervis*	1	46	9
	3	24	4		2	36	8
Polyscias fulva	1	184	24	*Chassalia subochreata*	1	18	1.5
	2	223	30		2	16	2
Lindackeria kivuensis	1	18	5	*Alchornea hirtella*	1	16	4.5
	2	16	4		2	15	3
Chassalia subochreata	1	16	5		3	20	5
Sapium ellipticum	1	30	29		4	23	8
Shirakiopsis elliptica	1	36	4		5	20	5

	2	153	28		6	17	5
Myrianthus holstii	1	18	4		7	21	5
	2	37	9		8	32	5
	3	63	14		9	20	5
	4	60	15		10	25	6
	5	45	6		11	21	4.5
	6	66	12		12	27	7
	7	77	10		13	17	3
	8	34	9		14	21	4
	9	18	5		15	15	4

	10	23	7		16	24	4.5
	11	33	8		17	16	3
	12	23	7		18	20	6
Syzygium parvifolium	1	103	18		19	17	3.5
	2	134	24		20	20	4
	3	78	21		21	15	4
	4	45	15		22	17	4.5
	5	24	5		23	15	4
	6	84	24		24	16	5
	7	17	5		25	16	3.5
	8	31	10		26	17	4.5
	9	18	9		27	15	5
Cyathea sp	1	33	4		28	18	5.5
	2	28	2		29	18	6
	3	23	2		30	16	4
	4	24	2		31	16	4
	5	30	1.5	*Chrysophyllum gorungosanum*	1	180	25
Rwahirirwa					2	144	26
Galiniera saxifraga	1	19	5		3	197	27
	2	34	10		4	59	8
	3	42	12		5	263	35
	4	30	8		6	272	25
	5	15	5	*Strombosia scheffleri*	1	114	25

	6	19	7		2	115	24
	7	28	9		3	88	20
	8	22	10		4	61	19
	9	38	8		5	71	35
	10	33	9	*Xymalos monospora*	1	75	9
	11	31	7		2	31	4
	12	23	3		3	30	4
	13	18	3		4	101	12
	14	17	5		5	16	4
	15	25	5	*Synsepalum attenuatum*	1	78	19
	16	22	4		2	202	35
	17	36	7		3	130	25
	18	36	9	*Magnistipula butayei*	1	23	4
	19	27	7	*Rauvolfia mannii*	1	16	3
	20	40	8	*Dracaena afromontana*	1	18	3
	21	21	2		2	16	3
	22	32	4		3	79	6
	23	40	8		4	21	2.5
	24	20	9		5	37	4
	25	41	10	*Psychotria palustris*	1	20	3
	26	17	7		2	18	4
	27	38	7	*Sericanthe burundensis*	1	18	4
	28	33	9		2	19	5
	29	42	9		3	16	5
	30	21	6	*Symphonia globulifera*	1	37	15
	31	36	10		2	19	3
	32	43	7		3	24	7
	33	39	8	Indét	1	121	18
	34	34	6	*Polyscias fulva*	1	219	29
	35	52	9	**Ndubura 3**			
	36	35	8	*Macaranga kilimandscharica*	1	121	20
	37	35	7		2	84	25
	38	42	6		3	27	9

Species	No.	Value 1	Value 2	Species	No.	Value 1	Value 2
	39	29	8		4	51	12
	40	20	5		5	69	12
Macaranga kilimandscharica	1	45	15		6	158	30
	2	67	17	*Chrysophyllum gorungosanum*	1	192	22
	3	51	17		2	39	12
	4	59	14		3	165	26
	5	22	9		4	156	22
	6	30	9		5	37	6
	7	27	11		6	142	25
	8	33	13		7	55	20
	9	64	18		8	133	20
	10	65	18		9	98	35
	11	21	11		10	29	2
	12	39	13		11	76	20
	13	82	15		12	88	24
	14	20	18		13	87	15
	15	29	12	*Myrianthus holstii*	1	15	4
	16	64	19		2	71	12
	17	33	10		3	59	12
	18	55	18		4	44	11
	19	61	14		5	33	7
	20	49	12		6	50	12
	21	58	15		7	24	16
	22	36	12		8	257	25
	23	40	14		9	31	9
	24	45	15		10	25	7
	25	57	16		11	18	4
	26	74	18		12	122	20
	27	38	15		13	64	15
	28	95	15		14	39	12
	29	30	11		15	25	7
	30	34	10	*Strombosia scheffleri*	1	22	7
	31	45	13		2	78	15

Species	No.	Value 1	Value 2	Species	No.	Value 1	Value 2
	32	43	13		3	187	36
Xymalos monospora	1	56	15		4	106	20
	2	47	16.5		5	27	9
	3	35	8		6	16	5
	4	70	13		7	31	6
	5	46	8		8	21	3
	6	17	3		9	32	7
	7	35	7		10	28	7
	8	23	6		11	22	0
	9	32	9		12	36	11
	10	52	7		13	21	5
	11	54	11		14	17	7
	12	65	12		15	90	20
	13	47	7		16	81	25
	14	26	3	*Erythrococca bongensis*	1	28	5
	15	16	9	*Psychotria palustris*	1	19	4
	16	42	8		2	16	3
	17	42	10.5		3	174	20
Syzygium parvifolium	1	145	18	*Rytigynia bridsonii*	1	19	4
	2	35	18		2	19	9
	3	51	18	*Garcinia volkensii*	1	15	4
	4	105	13		2	15	4.5

	5	21	8
	6	209	19
	7	142	18.5
	8	94	35
	9	268	19
Psychotria palustris	1	15	4
	2	15	2
	3	17	10
	4	32	7
	5	23	6
	6	15	4

3	22	5
4	23	5
5	18	6
6	16	17
7	15	4
8	25	6
9	17	4
10	16	7
11	21	6
12	15	5
13	27	2.5

	7	22	3
Dracaena afromontana	1	68	10
Indét	1	166	19
Tabernaemontana johnstonii	1	68	13
Symphonia globulifera	1	53	18
	2	27	11
Maesa lanceolata	1	51	11
Myrianthus holstii	1	25	7
Bersama abyssinica	1	61	9
Cyathea maniana	1	34	5
	2	35	4.5
	3	27	2
	4	28	2.5
	5	37	6
	6	28	5
	7	38	3.5
	8	34	3.5
	9	34	4
	10	37	2.5
	11	34	4.5
	12	37	4
	13	35	3.5
	14	18	4
	15	35	3.5
	16	28	2.5
	17	35	3
	18	31	2.5
	19	32	1.5
	20	32	1
	21	28	1
	22	26	3
	23	36	1.5
	24	32	2

	14	18	8
	15	15	7
Tabernaemontana johnstonii	1	77	15
	2	115	14
	3	23	9
	4	107	15
	5	71	18
	6	25	6
	7	34	10
	8	89	12
	9	126	25
	10	42	12
	11	42	10
	12	97	12
	13	24	7
	14	104	17
Dracaena afromontana	1	22	6
	2	17	5
	3	56	8
	4	24	5
Symphonia globulifera	1	308	34
	2	27	6
	3	54	25
Maesa lanceolata	1	39	12
Xymalos monospora	1	76	25
Neoboutonia macrocalyx	1	53	18
Magnistipula butayei	1	16	4
Syzygium parvifolium	1	144	25

	25	34	3
	26	36	4
	27	38	3.5
	28	41	3
	29	32	1.5
	30	38	2.5
	31	35	3
	32	39	2
	33	33	5
	34	35	1.5

1: *Maesa lanceolata*; 2: *Myrianthus holstii*; 3: *Symphonia globulifera*; 4: *Syzygium parvifolium*; 5: *Strombosia scheffleri*; 6: *Dracaena afromontana*; 7: *Chrysophyllum gorungosanum*; 8: *Synsepalum seretii*; 9: *Xymalos monospora*; 10: *Polyscias fulva*; 11: *Macaranga kilimandscharica*; 12: *Dombeya goetzenii*; 13: *Neoboutonia macrocalyx*; 14: *Prunus africana*; 15: *Tabernaemontana johnstonii*; 16: *Phytolacca dodecandra*.

17 18 19 20

21 22 23 24

25 26 27 28

29 30 31 32

17: *Crassocephalum montuosum*; 18: *Ensete ventricosum*; 19: *Schefflera goetzenii*; 20: *Smilax anceps*; 21: *Psychotria palustris*; 22: *Galiniera saxifraga*; 23: *Gynura scandens*; 24: Flowers of *Gynura scandens*; 25: *Virectaria major*; 26: *Triumfetta cordifolia*; 27: *Piper capense*; 28: *Setaria megaphylla*; 29: *Sericostachys scandens*; 30: *Discopodium penninervium*; 31: *Garcinia volkensii*; 32: *Coccinia mildbraedii*.

33 34 35 36

37 38 39 40

41 42 43 44

45 46 47 48

33: *Acalypha ornata*; 34: *Tacazzea apiculata*; 35: *Canarina eminii*; 36: *Dalbergia lacteal*; 37: *Rumex abyssinicus*; 38: *Monanthotaxis orophila*; 39: *Isachne mauritiana*; 40:*Hibiscus ludwigii* with flower; 41:*Urera hypselodendron*; 42: *Begonia meyeri-johannis* ; 43: *Brillantaisia cicatricosa*; 44: *Chassalia subochreata*; 45: *Schefflera abyssinica*; 46: *Stephania abyssinica* ; 47: *Momordica foetida* ; 48: *Vepris renieri*.

49 :*Bersama abyssinica*; 50: *Solanum nigrum*; 51: *Kalanchoe crenata*; 52:
Laportea alatipes; 53: *Brachystephanus africanus*; 54: *Desmodium repandum*; 55:
Jaundea pinnata; 56: *Arisaema mildbraedii*; 57: *Cyphostemma mildbraedii*; 58:
Mimulopsis solmsii; 59: *Parinari excelsa*; 60: *Gymnosporia acuminata*; 61:
Pteridium aquilinum; 62: *Mikaniopsis usambarensis*; 63: *Clerodendrum* Sp.;

Printed by Books on Demand GmbH, Norderstedt / Germany